The Complete Book of Home Plumbing

The Complete Book of Home Plumbing

PETER JONES

Art by Ralph Haarup

Charles Scribner's Sons
New York

Library of Congress Cataloging in Publication Data

Jones, Peter, 1934–
    The complete book of home plumbing.

    Includes index.
    1. Plumbing—Amateurs' manuals.
I. Title.
TH612.J66   696'.1   79-67071
ISBN 0-684-16298-9

1 3 5 7 9 11 13 15 17 19 V/C 20 18 16 14 12 10 8 6 4 2

Printed in the United States of America

Contents

Introduction

Plumbing is easy. It looks complicated because of the size of the pipes and the strange directions they take as they consume all the space under your sinks and stretch across your basement ceiling. But plumbing is still easy and the only tool you absolutely need to have for most plumbing work is a $6 pipe wrench.

You can read this book from cover to cover and memorize everything you need to know about plumbing. Then you can begin to disassemble and reassemble all the plumbing in your house, or better still, practice on a neighbor's house, and quickly develop all the manual dexterity needed to complete any plumbing chore. But most people will not be able to remember everything they read between these covers, particularly if they have no occasion to use any of the information for a time. And when the situation arises and you need to reread a portion of the book, it may not be handy. But there are some very fundamental truths that you can bear in mind about plumbing that will serve you in good stead no matter what you are doing.

When you pick up your tools and proceed to work on the house plumbing you can be certain of these facts:

1. Some pipes are threaded so that a fitting can be tightened on them by turning it clockwise. It can be loosened by rotating the fitting coun-terclockwise. This is true for 98 percent of all pipes and fittings.

2. The threads on the end of a pipe are conical; they are narrower at the tip of the pipe and gradually become larger and larger. You can split a fitting just by tightening it around a pipe thread too much. So only tighten as far as you can go without straining mightily.

3. The drain hole in the bottom of sinks and bathtubs is only a hole. It has no threads in it. There will be times when you will think it is threaded when you are trying to remove an old sink strainer and it refuses to budge no matter what you do. Whatever you do to get the strainer out of the fixture, you do not have to unscrew it.

4. All drainage pipes must slope at least ⅛ inch for every foot of their run toward the soil stack. The only exception is the drainpipe on appliances such as a waste disposer, washing machine, or dishwasher. Because these units pump out their waste water, the drain slope is not mandatory.

5. All pipe connections must be watertight. Very often a leaking connection can be fixed merely by turning the fitting a quarter or half a turn tighter. (See Number 1 above.)

6. All faucet handles are held to their stems by a screw. The screw may be hidden under a decal that must be pried loose, but there is a screw.

7. All globe-type valves, no matter what they look like, function and can be repaired exactly in the same manner as any standard faucet. The only difference between a globe-type valve and a globe-type faucet is that faucets have spouts.

8. You cannot connect pipes made of dissimilar materials without using special fittings. In other words, if you are installing plastic tubing and must connect it to a galvanized steel pipe, you must use an adapter that can be threaded onto the metal at one end, and solvent-welded to the plastic at the other.

9. All pipes should go as directly to their destination as possible. They may look as if they do not always do that, but they should.

10. All faucets must have an air cushion in their water supply piping.

11. All faucets should have a shutoff valve as close to them as possible.

12. All fixtures must have traps in their drain lines.

13. In times of emergency, when water is pouring through the walls and dripping off the ceiling, shut off the nearest supply valve, or turn off the main water supply valve. The main valve is usually positioned at the water service entrance. Be sure every member of your household knows where it is and how to close it.

14. When dealing with any plumbing, persevere. Be assured that no matter what you are trying to do there is probably a fitting or a tool that will allow you to do it. Your best chance of finding that supply or tool is at a plumbing supply outlet listed in the yellow pages of your local telephone book.

If you have the right tools and the proper supplies, plumbing is easy.

The Complete Book of
Home Plumbing

1 The House Plumbing System

There are some basic tenets of plumbing that anyone approaching his or her water system should fully understand. To begin with, for some reason known only to the Almighty, plumbing problems most often arise at such inopportune moments as the middle of the night or during a national holiday when it is impossible to find a plumber, let alone one who is available for repair work. Secondly, if you do find a plumber, he is expensive, particularly in light of the simple fact that 90 percent of all plumbing repairs made in the home can be performed by any man, woman, or child over the age of ten.

If you go down to your basement and stare up at the ceiling, or if you look under your sink, admittedly the maze of pipes and valves appears to be very formidable indeed. But there is no mystery to them once you understand that every pipe is going about its business as directly as it can; and its business is to lead from, or go toward, a source of water. Nor is there anything complicated about assembling all those pipes. In fact, the only secret to any kind of plumbing is to use the proper tools, which are neither numerous nor particularly expensive. Actually, with the exception of a 10-inch pipe wrench costing about $6, you don't even have to buy any of those tools until the need arises.

As to the mysteries of proper plumbing, there is only one: No matter what the purpose of any pipe, no matter which way it twists or turns, it will always function to perfection so long as it is neither clogged nor leaks. In practical terms that means every joint where two pipes come together must be made with a watertight seal.

THE MAIN PLUMBING SYSTEMS

The majority of all repairs in any home system occur somewhere around the fixtures—the sinks, toilets, or tubs—and nearly every one of those repairs involves a joint that is either leaking water or a pipe that does not permit water to pass through it. However, it is helpful to understand your entire plumbing system and how it works before you attack any repairs.

Every home plumbing system divides into three separate but interconnected systems: the water supply lines, the fixtures, and the drain-waste-vent (DWV) system. The *water supply pipes* bring fresh water from your well, storage tank, reservoir, or a municipal water supply system into the house and distribute it to each of the fixtures. The *fixtures* are all the outlets used for getting water out of the water supply pipes and include basins, sinks, tubs, showers, toilets, and lavatories. The *drain-waste-vent* system is a separate set of pipes that lead away from each fixture and carry

1

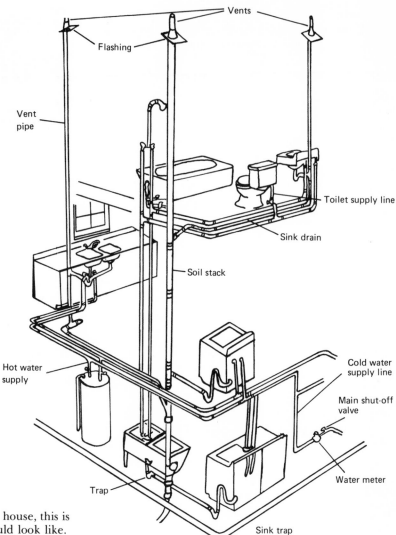

If you could see through the walls of your house, this is
what your complete plumbing system would look like.

all waste and used water to a septic tank, cesspool,
or the municipal sewage system.

Water Supply System

The fresh water that enters your house can come
from any of several sources, including a well,
stream, reservoir, storage tank, or municipal water
system. Water is brought into your house through a
single, 1-inch diameter brass, copper, or lead pipe,
and it arrives under as much as 100 pounds per
square inch (psi). Immediately inside the house

there is a main shutoff valve that can close off all
the water entering the building. If the water is
metered, a meter is normally located just past the
main valve, although in many communities the
meter may be attached to the outside of the building. It may even be located in a miniature manhole
somewhere on the lawn, which can be opened so
that the meter is accessible for inspection.

As the water leaves the water meter, or at least
once it is past the main house valve, there may be a
pressure-reducing valve, which is designed to reduce the pressure of the water, probably to about
60 psi. A single pipe filled with cold water leads

away from the meter or pressure-reducing valve and begins to course its way through the building, carrying cold water to every fixture in the house. One of the first fixtures it supplies is the hot water heater. The heater warms the water to a temperature of between 140°–180°F and sends it into the hot water main, which begins at the heater and also runs through the house to each of the fixtures. Normally, the hot and cold water mains are parallel to each other wherever they go and are usually ¾-inch diameter pipes; although they may be as large as 1 inch or as small as ½ inch in diameter. Any pipes that branch off the mains and lead directly to the various fixtures may be as small as ⅜ inch in diameter.

At each fixture the cold water line is connected to the right faucet and the hot water line to the left. In a properly assembled water supply system both the hot and cold water lines are also provided with an air chamber as close to each faucet as possible. The *air chamber* is merely a 12-inch to 18-inch length of pipe that extends vertically up from the water line and is capped. It may be inside the wall behind the fixture, or you may be able to see it somewhere near the faucet; its purpose is to prevent the rattle known as water hammer. Water cannot be compressed, so whenever a faucet is turned off, the running water comes to an abrupt halt and rattles the pipes. The air chamber provides a column of trapped air, which acts as a cushion against the several hundred pounds of water pressure suddenly exerted on the supply lines.

Throughout the water supply system are a number of valves used to shut off the water to various parts of the house. A good plumbing system includes a shutoff valve everywhere the hot and cold water lines emerge from the wall near a fixture. You will also find valves in the basement that allow you to turn off the water to a single fixture such as a dishwasher or clothes washer. Still other valves are positioned so that entire sections of the plumbing system can be shut off without affecting the water service to other parts of the house.

The pipes used for the water supply system are often made of galvanized steel or copper. In many parts of the United States plastic pipe specially manufactured to withstand temperatures of more than 180°F may also be used. In a few areas with "problem" water, brass or plastic piping is commonly used.

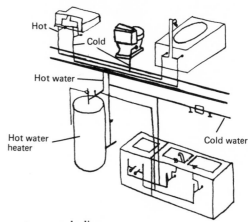

The water supply lines.

Fixtures

The most obvious elements in any plumbing system are its fixtures—the bathtubs, sinks, lavatories, showers, and toilets. No matter what its design, every fixture consists of at least one faucet, which is connected to the water supply system, and a receptacle, which can hold the water but which is also connected to the DWV system.

Toilets are somewhat unique because they use only cold water, which is handled automatically by a float valve inside the toilet tank. The only control you have over a toilet involves the discharge of its water.

Drain-Waste-Vent System

The drain-waste-vent (DWV) system is made up of vents, traps, waste pipes, and a soil stack, all of which are designed to carry used water and waste away from the fixtures.

Between every fixture and the DWV system there is a curved pipe called a *trap*. Traps are designed to hold water in the bottom of their curves after a sink or a bathtub is drained, and thus prevent gases and vermin from rising up the pipes and entering the house. Sinks and lavatories usually incorporate a P-shaped or S-shaped trap, while toilet bowls are designed to function as their own traps by retaining water at all times. There are three kinds of traps found in home systems. P and S traps are simply 1¼-inch or 1½-inch-diameter pipes shaped in the form of the letters P or S. The

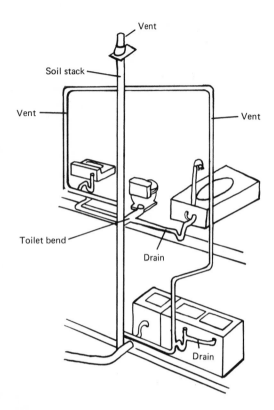

The DWV system.

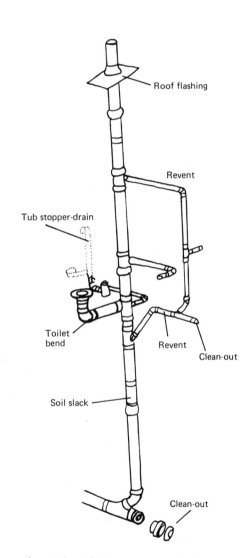

The main stack with its attendant drains and vents.

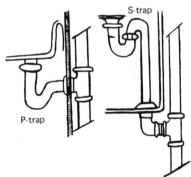

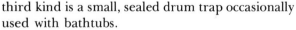

P and S traps.

third kind is a small, sealed drum trap occasionally used with bathtubs.

The pipe leading out of the trap is called the *waste arm* and it goes directly to the fixture waste pipe, which almost always has either 1¼ inch or 1½ inch in inside diameter. The fixture waste pipe is always sloped so that waste water will flow away from the fixture toward a 3-inch or 4-inch vertical pipe called the soil, or main, stack. It should be

noted that all horizontal pipes in the DWV system are always sloped approximately ¼ inch for every foot of their run.

The soil or main stack is a vertical pipe that goes from the building drain at the bottom of the house to the roof, where it is capped by an open vent. The soil stack has two purposes: 1) to let air into the plumbing system and thus keep the traps from being siphoned dry every time water rushes through them; and 2) to allow sewer gases to escape without putting pressure on the seals in the traps. In older houses the stack is invariably a cast-iron pipe 4 inches in diameter, but many new homes use either a copper or plastic pipe. All waste and used water from the fixtures runs down the soil stack, then along a 3-inch or a 4-inch pipe at the bottom of the house known as the *building drain.*

The building drain is also pitched so that waste can flow to the street sewer, cesspool, or septic tank. It is fitted with a capped cleanout opening at the beginning of its run and again just before it leaves the house, in case the pipe ever becomes clogged.

Cleanout plugs are normally located at the base of the stack as well, and anywhere else the waste line changes direction. In cast-iron mains the cleanout plug is often brass, and after several years it may "weld" itself to the iron and become impossible to unscrew. The fitting must then be cut out with a cold chisel and replaced by a lead or plastic plug that will not "freeze" shut.

Some houses also have a U-shaped trap located just before the building drain leaves the house.

The usual material for drainpipes is 4-inch-diameter cast iron, primarily because it will last in excess of 40 years once it is installed. However, cast iron is difficult to work with and as a result 3-inch copper drains are often found in modern homes, as well as 3-inch PVC and ABS plastic pipe. Galvanized steel pipe also shows up as part of many drainage systems, particularly at the points where fixtures are connected to the rest of the DWV system.

A properly vented plumbing system allows pressure in the supply pipes to remain equalized so that waste and water can flow freely, and no water can back up into any fixture stationed below a draining fixture.

Thus, every trap in any plumbing system should be vented directly to an open stack that extends up through the roof of the house. Technically, the section of any stack immediately above a fixture is designated as a vent; all pipe below the fixture is known as the waste line. If a trap cannot vent directly outside, it should have a vent pipe called a *revent* that runs at right angles to, and connects with, a stack *above* the highest fixture connected to that stack.

Any part of a fixture's waste pipe that also serves as a vent for itself or another fixture is known as a *wet vent.* The length of a wet vent is determined by the diameter of the pipe. The longer the run, the larger the vent pipe must be so that it can never completely fill with waste water that would cause suction and therefore pressure on the trap seals. In larger houses one main stack is often not enough. Some fixtures are therefore connected to smaller stacks.

THE HIGH COST OF PLUMBING REPAIR

The cost of hiring a plumber ranges between $15 and $30 an hour plus parts and travel time, depending on where you live and whether you hire an individual plumber or a contracting company. At those rates it pays to do as much of your own plumbing as you can; for example, it hardly makes economic sense to pay a plumber $15 or $20 to replace a 5¢ faucet washer when almost any member of your family could do the job in less than half an hour.

There are, of course, some plumbing tasks that demand special tools or that are just too big for you to take on in your spare time. But the fact remains, that the majority of breakdowns that occur in any plumbing system are simple to repair.

PLUMBING CODES

Like electrical wiring, major plumbing in your home must comply with local health and safety codes. So before you tackle any major plumbing, such as adding a bathroom to your house, get a copy of the city or county building department regulations that govern plumbing in your area. Many local codes require an officially witnessed pressure test of any new water supply lines that are installed. Other municipalities require a building department permit for any plumbing that is modified (as opposed to repaired). Repairs do not require a permit. Or your work may have to withstand an official inspection if you do any major plumbing work. Many locales also have specific restrictions on the kinds of pipes that can be used, particularly in the case of plastic pipe. Versatile as it is, plastic piping has yet to achieve the complete acceptance of every community.

Primarily, the plumbing codes are most concerned that your plumbing system does not have any *cross-connections* between the water supply lines and the DWV, which can cause fresh water to become contaminated, thereby creating a health hazard. A cross-connection is a direct hookup between the potable water system and any source of possibly contaminated water. One of the most common cross-connections is the submerged inlet of an old-fashioned sink or bathtub. Should the bowl be full and a backflow of water occur, as it might during a loss of city water pressure, possibly contaminated water from the bowl could be drawn backward into the house water supply system. Other common cross-connections are garden hoses left submerged in water, boiler-fill pipes that are not equipped with backflow-prevention devices, and laundry tub hoses left connected with their ends submerged in water.

To prevent the dangers of cross-connections modern fixtures are all built with air gaps between the flood rim and the lowest faucet discharge opening. The most up-to-date plumbing codes now require that every household have an effective device at the water service entrance to protect others in the area from risk of disease and possibly death from drinking back-flowed water.

Plumbing codes are, if nothing else, an excellent guide to efficient, sanitary plumbing systems; and they should be adhered to closely.

2 The Plumber's Tool Kit

The majority of plumbing repairs in your home will require a pipe wrench, an open end wrench, a Phillips head and a regular slot screwdriver, a suction force cup (plunger), and a pair of pliers. You could go for years and never need any other tools unless you decided to build a new bathroom. There are, however, several other tools that plumbers use and that you may someday have occasion to need. Here is a breakdown of the most usual ones.

WRENCHES

Adjustable Open-End Wrench

This wrench has one fixed jaw and one that slides along a screw to adjust the opening. The jaw faces are smooth; therefore they are useful for gripping bonnet nuts on faucets and other fixtures but not pipes. Adjustable open-end wrenches can be purchased in many sizes, with their jaw opening getting wider as the tool becomes longer. The 12-inch wrench is the most useful for plumbing repairs because the jaws will open to more than an inch, which is wide enough for gripping most bonnet nuts.

Pipe Wrench

For turning and holding pipe, the adjustable pipe wrench is an absolute necessity. In fact, you should own two of them. The pipe wrench has serrated jaws, one of which is movable, and a pair of pipe wrenches, one 12 inches and one 10 inches, will let you handle practically any pipe or pipe fitting in your home. Pipe wrenches are designated by their overall length, but it is the width of their jaw opening that really counts. A 10-inch pipe wrench will open to 1¾ inches, which is enough to get at most drainpipe fittings.

Basin Wrench

If you are disconnecting or installing a sink, you will find this tool indispensable. It consists of a long handle on the end of which is a serrated, adjustable jaw that can grip the fittings that connect the faucets to the water supply lines. Given the confined spaces where they are usually hidden, sink nuts are almost impossible to get at with other types of wrenches.

Chain and Strap Wrenches

These are used for handling large-diameter pipe, and they will not mar a chrome or brass finish. You will probably never need to own one.

Open-End Nut Wrenches

You can buy these separately or in sets; a set consisting of sizes from ¼ inch to ¾ inch is sufficient for most plumbing chores. The chores will

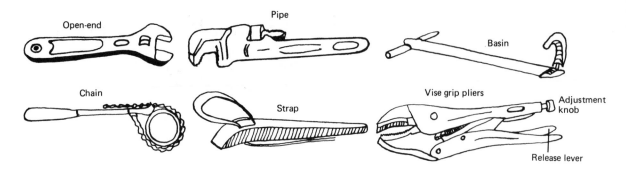

Wrenches used in plumbing.

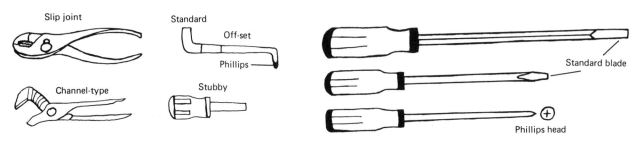

Pliers and screwdrivers

consist of removing small lock nuts, faucet bonnets, etc., from areas that are inaccessible to larger tools.

Vise-Grip Wrenches

These look like ordinary pliers but they have the curved, serrated jaws of a pipe wrench. They can be purchased in several sizes and if you have one that is big enough, it may be used on pipes and fittings in the same manner as a pipe wrench. A popular size is 10 inches.

What makes the vise-grip wrench unique is its adjustable screw in the handle, which widens or closes the jaws. Once the jaws are fitted around an object, the lever handle is squeezed shut and the tool locks onto it like an angry fox terrier. It locks so well that most models also have a releasing lever that can be pulled to open the jaws.

PLIERS AND SCREWDRIVERS
Pliers

A good set of common slip-joint pliers, including square-nosed as well as long-nosed pliers, can be very handy to have when you are working with small nuts, fittings, and so on, especially when you find yourself in a confined area. You can also buy a pair of *channel-lock* pliers, which have an extra-wide grip that is extremely useful as an alternative to using a pipe wrench.

Screwdrivers

You should have in your possession a set of screwdrivers that includes two or three blade widths in the standard versions, and at least two different diameters of Phillips head drivers. If you buy a set of, say, seven screwdrivers, it will probably also include an S-shaped offset screwdriver with a Phillips head at one end and a standard blade on the other. It looks weird, but you will find yourself reaching for it with surprising regularity because it can be used in tight places.

The most important thing to know about any screwdriver is that its blade should fit exactly into the slot of the screw. If it is too large, it will not get a decent grip on the screw. If it is too small, it will slip and gouge the edges of the slot until the screw becomes impossible to turn.

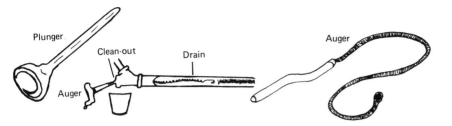

Plungers, snakes, and augers

PLUNGERS, SNAKES, AND AUGERS

Rubber Force-Cup

Call it whatever you wish—"the plumber's friend," or plunger—you can't live without one. Anytime a sink, lavatory, bathtub, or toilet becomes clogged, a plunger is your first resort. There are several types available; but buy the kind with a retractable bulb. When you use it, the bulb should fit snugly into the bowl opening to get the best possible suction. Then, push down and pull up again until the obstruction is cleared. Buy a good-quality, heavy-duty plunger with a 6-inch cup.

Plumbing Snakes

There are various types and sizes of these available on the market, but the two most popular versions are the coil spring and flat ribbon types. Both of them will do the job of freeing a clogged drain. With any snake, you provide the power. You insert the snake into a drain and jab, push, twist, and strain until the obstruction comes free and the line is clear again.

Closet (toilet) Auger

A close relative of the snake, this is designed to unstop a toilet bowl. Basically, it consists of a coil-spring snake plus a turning handle that makes the auger function much like a woodworking brace and bit in that it literally drills through the obstruction.

If you possess a representative of each of the tools listed so far, you can assume you are equipped to handle any plumbing emergency and most repairs in your home. But if you are contemplating some plumbing renovations that will require adding more pipes to your system there are a few other tools you should consider. Surprisingly, most of them are not very expensive.

HAMMERS, COLD CHISELS, AND SAWS

Ball Peen Hammer

A ball peen looks like a regular carpenter's hammer, except instead of a claw for pulling nails it has a ball behind the back of its flat head. The weight most people use is 16 ounces, which is the weight carpenters usually pick for their claw hammers. What the ball gives you is the ability to hammer out metal without making a lot of flat dents. Most often the ball peen hammer is used for hammering on steel cold chisels and caulking tools, but be careful with it when you are working with copper tubing or brass fittings. It will dent them both with no trouble at all.

Cold Chisels

If you get into working with cast-iron soil lines, you will need some cold chisels of varying widths for cutting the pipe. They are also useful for breaking up mortar and bricks, and cleaning away large accumulations of rust or sludge from around large drains.

Saws

You may, on occasion, need the various carpenter's saws when doing roughing-in work for plumbing

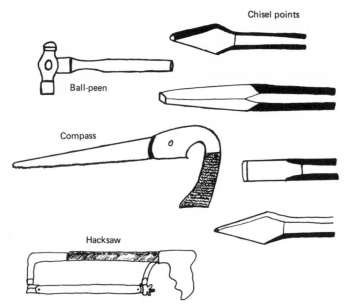

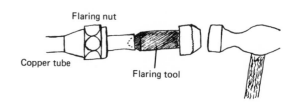

Pipe and tubing cutters are identical except for the thickness of their cutting wheels.

Flaring tools are both inexpensive and easy to use.

The ball peen hammer, cold chisels, and saws.

lines. You will specifically find that a $2 utility saw is invaluable. They usually come with both a wood and a hacksaw blade and are essential when you are working in tight places. A standard *hacksaw* is an absolute necessity when you are cutting pipe or tubing, particularly in places where a pipe or tube cutter cannot be used. As a plumber you will practically always be working with thin-walled piping or tubing and will therefore need a 10-inch to 12-inch blade with between 24 and 32 teeth per inch.

OTHER PLUMBING TOOLS AND SUPPLIES

Pipe Cutters

Get a one-wheel, two-roller type that can be adjusted to pipes from ½ inch to 2 inches. The cutter consists of a vise, a cutting wheel, and a handle to tighten the cutter against the pipe. You put the pipe in the vise and tighten the cutter against the metal, then rotate the pipe cutter until it turns smoothly. Tighten the handle again and apply a few drops of oil to the cut. Keep tightening and rotating the cutter until the pipe is severed.

Tubing Cutters

These look like pipe cutters and are used in the exact same manner. Their only difference is they have a thinner cutting wheel and they are supposed to be used only on copper or brass tubing. You will need one that can cut between ¼-inch and ¾-inch diameter tubing. There is a little metal arrow that protrudes from the back of the vise on the tubing cutter, called a reamer. Once you have cut a piece of tubing, stick the arrow into the end of the tube and rotate it to clean the burrs off the inside surface of the metal.

Files and Rasps

You can get these separately or in sets. No matter how you buy them you have a wide range of teeth and cutting edges to choose from, but be careful to purchase the kind used for metal, not wood. You ought to have at least one fine, one medium, and one coarse version, as well as a stock of steel wool and emery paper. All of these are used for getting rid of unwanted roughness on the ends of pipes you are cutting and fitting.

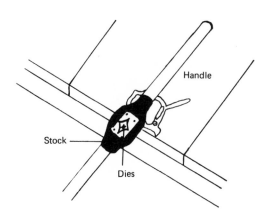

Handle

Stock

Dies

You can spend well over $100 for a stock and dies set, so be sure you have a lot of pipe threading to do before you make the investment.

Flaring Tools

Flaring tools are used to spread the ends of tubing so that they will make a watertight joint. The tools amount to a small vise with varying sized holes, known as a slip-on yoke, and a conical, screw-driven flare point. The most versatile size to have is one that allows you to flare tubing from ⅛ inch to ¾ inch in diameter. You put the pipe in its proper hole size in the yoke and tighten the wing nuts to secure it. Then slide the yoke until it is directly under the flaring point. Now screw the flaring point into the metal until the end of the tube is flattened against the yoke. When that is complete, take the tube out of the yoke and flare the end of the tube that is to match it. If you do it properly, flared tubes and their fittings will produce a virtually watertight seal. Be sure to install the flare nut before making a flare. Otherwise you will not be able to get it on the pipe.

Stock and Dies

These are used to produce a thread on the outside ends of pipes so that they can screw into pre-threaded fittings. If you are going to buy a stock and die set, get one that allows you to thread pipe from ½ inch to 2 inches in diameter. The procedure for using a stock and die set is simplicity itself:

Place the pipe to be threaded in a vise, leaving about a foot of pipe extending beyond the face of

the vise so you have plenty of room for turning the stock and die. Insert the proper diameter die in the stock and slip the stock squarely over the end of the pipe. Push it in until the die touches the front edge of the pipe. Keep the stock absolutely square with the pipe and begin turning it clockwise. You will have to stop from time to time and apply cutting oil to the metal; oil protects the die, keeps the threads from tearing, and makes your work easier. Keep oiling and rotating the stock clockwise until you can see approximately ¹/₁₆ inch of threads beyond the die, then rotate the stock counterclockwise, backing it off the end of the pipe. You will have to clean the newly cut threads to get rid of any particles of metal in them.

Propane Torch

This is a bottled gas unit with a flame control valve and it comes in kits that include a starter, extra gas tank, and different flame points. You cannot "sweat" any copper or brass joints, or do any plumbing soldering without one. Follow the manufacturer's instructions exactly when using one.

Seat Dresser

The purpose of the seat dresser, or seat grinder (as it is commonly called), is to resurface damaged valve seats so that the seat washer will fit properly. Seat dressers look and act somewhat like a corkscrew and you can buy them in different styles and with various size cutting heads. To use a seat dresser you must first remove all the parts of the valve or faucet until you can look down into the valve seat. Now insert the cutter of the seat dresser into the valve body and start rotating its handle. Keep turning slowly, taking small cuts at a time until all the scars and burrs have been removed from the seat. Then reverse the handle, thereby disengaging the cutter, and remove the dresser. You will have to flush out all of the particles left in the valve seat before reassembling the faucet.

Pipe Vises

There are numerous types of pipe vises, including the combination vise you may already have attached to your workshop bench. The distinguishing feature of a pipe vise is that it has serrated jaws

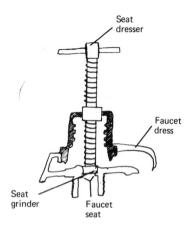

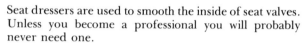

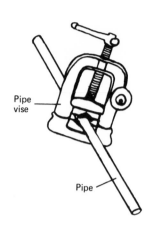

Seat dressers are used to smooth the inside of seat valves. Unless you become a professional you will probably never need one.

Pipe vises are distinguished by their curved, serrated teeth.

curved to receive round objects such as pipe. The standard pipe vise has a jaw width of 4 inches and can be swiveled or is movable. You can also find *chain vises* that can handle pipe diameters of ⅛ inch to 8 inches. But unless you are doing a considerable amount of big pipe work, you can get along with almost any type of pipe vise so long as it is large enough to accept a 4-inch pipe in its jaws. Beware when working copper, brass, or plastic pipe that you do not overtighten the vise. The primary function of a vise is to hold the pipe securely while it is being cut, threaded, or flared. If you overtighten it, you may rupture or distort the shape of the pipe or tubing.

Sealers

Any pipe joint that is not soldered together must have some kind of sealer applied to the connection to guarantee a watertight seal. Until a few years ago that meant putty, oakum, or a compound. But today, with the exception of a few old-time plumb-

ers who prefer not to change their ways, practically everyone uses plastic tape to seal their pipe joints. The white plastic tape (sold as teflon or TFE tape) is dispensed from a spool much as transparent gummed tape is. You simply wrap the tape once or twice clockwise around the male threads of the pipe or fitting and then make your connection. It is quicker, neater, and simpler than applying any of the compounds; and it will seal the joint equally as well for just as long.

One caution about sealing joints: Never apply the sealer to the female threading. It will probably not seal very well and is guaranteeed to produce some blockage on the inside of the connection.

Oils Used with Plumbing

Have a can of good, medium weight sulfur-based cutting oil on hand. You will need to lubricate both pipe and tubing while it is being cut or threaded. You can also use the oil for cleaning rust off porcelain and metal.

3 Plumbing Repairs

For the most part, the repairs your home plumbing system requires will amount to curing a leaky faucet or unclogging a drain, although once in a great while you may have to deal with a broken pipe. It is always a little scary when a pipe ruptures and suddenly there is water seeping through the walls or dripping from a ceiling. It can happen anytime a pipe has frozen and split, or when the main sewer line becomes clogged, causing your sinks to overflow. No matter where the trouble is, first shut off the water valves nearest the problem. If it is quicker or easier, close the main water supply valve, the one located nearest the water meter or where the main service line comes into your house. Once there is no more water entering the pipes you have plenty of time to locate the trouble and at least make some emergency repairs.

TEMPORARY EMERGENCY REPAIRS

When disaster strikes, here are some immediate procedures that will get your plumbing system back into temporary working order. However, none of these repairs, with the exception of unfreezing pipes, should be considered permanent. The repairs suggested here should only be performed in an emergency and at the earliest possible moment you should go back and fix the problem permanently.

Frozen Pipes

Pipes that have been frozen are in imminent danger of splitting, so work on them carefully and with patience. If you have an electric heating cable, wrap it around the frozen pipe, plug it in, and leave it to do its job. You can also wrap a frozen pipe in insulation, which will eventually warm it enough for the water to begin flowing again.

The most laborious, but quickest, way of unfreezing a water supply line is to dip rags in hot water, wring them out, and wrap them around the pipe. Apply your hot rags starting at the faucet and work back toward the house service entrance, never the other way. The rags should be reheated as frequently as necessary until the pipes unblock. While you are working, keep all the faucets along the way open so that as the water melts it has someplace to flow. As it moves, it will melt all the rest of the ice and help shorten your work time.

Split Pipes

A pipe can split for several reasons. It may have frozen; it might have been slammed by something heavy; it could have a natural weakness that failed to withstand the normal pressure of water moving through it; it might just be very old or have corroded. No matter what caused the rupture, sooner or later the damaged section must be entirely re-

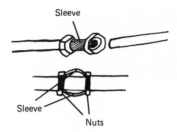

Assembling a dresser coupling.

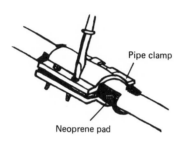

Assembling a compression clamp.

placed. But, provided the split is no more than 4 inches in length, you can make temporary repairs with either a *dresser coupling* or a *compression clamp*.

HOW TO USE A DRESSER COUPLING

Close the valve controlling the flow of water through the pipe and open the nearest faucet to drain as much water out of the pipe as you can. Cut the pipe with a hacksaw about 1 inch past either side of the damaged area. Do not take out more than 6 inches of pipe or the coupling will not have enough good pipe to clamp around.

Unscrew the compression collars at each end of the dresser-coupling body and slide one collar over each end of the severed pipe. Then insert the main body of the coupling over the cut pipe ends and push the collars over the ends of the coupling body. Tighten the couplings with a wrench until they are snug, then turn on the water and inspect the collars for leakage. If any leaks appear, tighten the couplings some more, until the water stops.

HOW TO USE A COMPRESSION CLAMP

Compression clamps are made of two curved pieces of metal lined with rubber or neoprene. The two halves are hinged on one side and have bolt holes on their opposite edges. But you have to have a clamp the same size as the pipe being repaired. Shut off the valve nearest the damaged part of the pipe and open the nearest faucets to drain the pipe. Then place the clamp around the damaged area; the clamp should reach at least 1 inch past each side of the split. Insert the bolts in the holes at the edge of the clamp, and tighten them alternately so that the two halves of the clamp come together evenly. Open the water supply valve; if the clamp leaks, tighten the bolts some more.

Minor Pipe Damage

If damage to a drainpipe is minor, such as a pinhole or hairline crack, you may be able to repair it with beeswax.

APPLYING BEESWAX

Heat the beeswax in a pan on your stove until it reaches the consistency of putty. Now force the beeswax into the pinhole or crack in the pipe and build it up over the damage. When the wax is cool and has hardened, wrap it with several layers of friction tape.

USING EPOXY

Epoxy is an alternative material for repairing minor damage in drainage pipes. First wipe the damaged area until it is absolutely dry. Sandpaper it to provide "tooth" for the repair. Then mix equal parts of epoxy resin and its hardener until they blend into a uniform color. Use a putty knife or your fingers to force the epoxy mix into the damaged area. Epoxy has a "pot" life of about a half-hour, so only mix as much as you can use within that time, and do not use the drain for at least 12 hours after the epoxy has been applied. By then, the epoxy will be rock-hard and the repair should last for some time.

REPAIRING FIXTURES

All the fixtures in your plumbing system are receptacles for holding water. They are serviced by fittings, that is, faucets, which bring water to them via the water supply lines; and they are attached to the DWV lines by drainpipes and traps. Since the

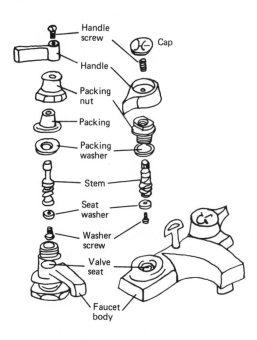

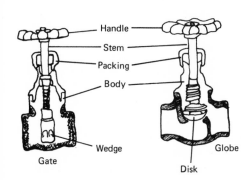

The anatomy of a gate valve (left) and a globe valve (right), both of which are commonly found in homes.

A standard faucet can appear in many forms but it always has the same basic parts.

faucets contain the only moving parts in a plumbing system, they are the most susceptible to leakage; so it is on the faucets that you will spend most of your time making repairs. At the same time, the drains and their attendant traps are designed to collect anything that is so large it might clog the DWV system. Consequently, the drains and traps are subject to occasional clogging, but look at it this way: at least they are more accessible than most of the DWV lines.

FAUCETS

If you look carefully, you will discover all kinds of faucets in your home, beginning with the valves that appear to be nothing more than handles sticking out of the water supply pipes. There are lavatory faucets, mixer faucets, and perhaps even the modern washerless faucets. Every faucet looks a little different, but essentially all of them function and are repaired in the same manner.

Most faucets and valves consist of a handle attached to a shaft that contains a rubber or plastic washer locked to its free end by a brass screw. When the faucet is closed, the washer presses against the bottom of the faucet body, so that water cannot pass it. When you turn the handle, the threaded shaft screws upward, lifting the washer so that water can flow out of the spigot.

The washer-type faucets are by far the most common versions in any home and they all have two problem areas: the seat washer and the packing nut. Nine times out of ten when a washer-type faucet drips, it is because the seat washer is worn and is not providing a watertight seal when the handle is closed.

Faucet Disassembly

While faucets vary considerably in both design and appearance, all of them are disassembled in about the same manner:

1. Turn off the water valve nearest the fixture.

2. Turn the faucet handle to its open position so that any water in the fixture can drain off.

3. Some fixtures have caps marked "Hot" and "Cold" over the handle screws. You will have to pry up the caps with a screwdriver. Then undo the screw in the top of the handle.

4. Pull the faucet handle upward from the stem shank. If you have any trouble with the handle, tap it gently and evenly around its bottom with the handle of the screwdriver. You might also try prying the handle *gently* upward with a screwdriver blade. *Do not force the handle.* It can break.

5. Wrap friction tape around the metal faucet bonnet to protect its finish. Now loosen the bonnet using an adjustable wrench.

6. Inspect the bonnet packing wrapped around the faucet stem. If the packing is compressed or worn it should be replaced with new bonnet packing, which you can purchase at most hardware stores. If there is no packing around the stem, look for a bonnet washer. If the washer is worn or chewed, or in any way less than perfect, replace it.

7. In order to remove the faucet stem, rotate it counterclockwise. You may be able to do this by hand. If not, put the handle back temporarily and use it to turn the stem.

8. Inspect the knurling around the inside top of the faucet body. Each tooth should be sharp, pointed, and not rounded. Also look at the threads around the stem to be sure they are sharp and not rounded. If either the threads or knurling is worn, the stem will have to be replaced.

9. The seat washer is held to the base of the stem assembly by a brass screw. Unthread the screw and inspect the washer for wear. If the washer is in any way damaged, replace it by undoing the screw and inserting a new washer in its place.

10. Inspect the faucet seat for any scarring. If you find that the seat is an integral part of the faucet body and is severely damaged you will have to replace the entire faucet assembly. If the seat is removable, you can unscrew it from the faucet body and replace it by simply screwing a new seat in its place. Sometimes you need a faucet seat tool. At other times a large screwdriver will get the job done.

11. After completing all inspections, begin reassembling the faucet by screwing the stem into the faucet. Tighten it snugly with a pipe wrench. But do not overtighten or you will split the metal. *The faucet stem should be left in a partially open position so that the stem washer does not bind in the faucet body.*

12. Place the packing washer over the stem shank, and make sure it is seated squarely. If the faucet requires bonnet packing, wrap the packing snugly around the stem.

13. Now place the bonnet on the stem and tighten it with your open-end wrench. Do not overtighten or you will risk stripping threads or compressing the packing.

14. Place the handle on the stem and replace the handle screw.

15. Turn the handle to the fully closed position and snap the handle cap into place.

16. Open the water valve and turn the handle on and off to test for leakage.

17. Remove the friction tape from the bonnet. Although any faucet can be disassembled following the above procedure, you will encounter some structural differences in the various faucet types. Here are the most notable variations:

WASHERLESS FAUCETS are much less vulnerable to leakage than standard types because they use, instead of a rubber washer, a revolving disk or diaphragm positioned inside the stem to stop the flow of water. Again, you are likely to encounter an insert (marked "Hot" or "Cold") in the top of the knob, which you have to pry off before you can expose the handle screw. Once you have the handle off you will not find either a packing nut or a washer, only the stem nut. Rotate the stem nut counterclockwise with a wrench and then lift the entire spindle assembly out of the faucet body. Washerless faucets may contain either a rubber diaphragm or rotating disks at their base. The disks have matching slots that line up when the handle is turned to the open position so that water can flow through the faucet. Since there are no washers these faucets seldom develop a leak; when they do, it is because of worn metal parts.

The washerless faucets sometimes have O-ring seals instead of packing. If the O ring fights you

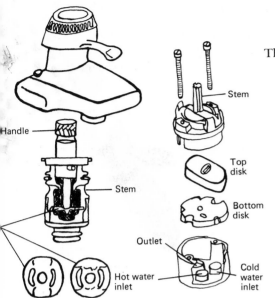

The two basic types of washerless faucets.

Handle

Stem

Stem

Top disk

Bottom disk

Outlet

Hot water inlet

Cold water inlet

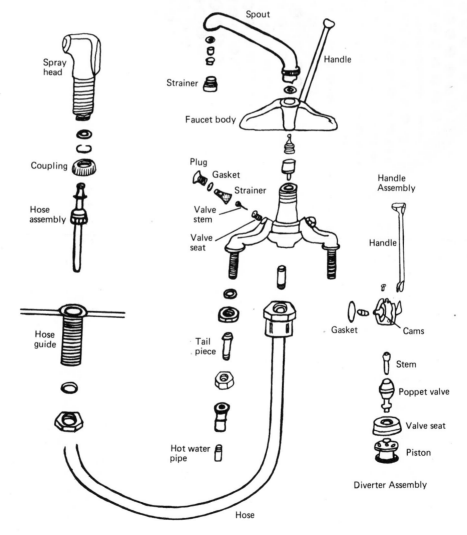

Anatomy of a single-handle mixer faucet and spray hose.

Spray head

Coupling

Hose assembly

Hose guide

Spout

Strainer

Faucet body

Handle

Plug

Gasket

Strainer

Valve stem

Valve seat

Handle Assembly

Handle

Gasket

Cams

Tail piece

Hot water pipe

Stem

Poppet valve

Valve seat

Piston

Diverter Assembly

Hose

when you are trying to remove the valve assembly, screw the handle back on the stem and use it as a grip to pull the assembly free. Do not use pliers or a wrench because either could damage the metal.

DISK VALVE FAUCETS

Disk valve faucets are another version of the washerless faucet. When you get the handle off one of them you will discover long bolts located on each side of the handle stem. These hold the unit together, and as soon as you unscrew them you can pull off the top half of the faucet. There is a movable disk with holes in it that rotates with the handle stem to open and close holes in a second disk fixed to the bottom of the faucet body. The fixed disk must be lifted out of the body to get at both the hot and cold water inlet seals. The metal parts in a disk valve faucet wear out occasionally and the inlet seals will deteriorate in time and need to be replaced.

SINGLE-HANDLE MIXER FAUCETS

There are numerous kinds of single-handle faucets, all having a single lever that controls both the hot and cold water. The kitchen sink version of single-handle faucets incorporates a spray hose attachment that occasionally malfunctions. But to work on one, you need a copy of the repair sheet that came with it. If you do not have one, write to the manufacturer.

SINK SPRAY HOSES

When a sink spray hose stops working, start by unscrewing the aerators in the faucet and the spray head and make sure they are not clogged. If the aerators cannot be cleaned or if they are damaged, replace them. Also examine the spray hose for kinks, clogging, or splits. If the hose is damaged in any way, unscrew its hex nut connector at the base of the faucet, and replace it. If all of these repairs fail, remove the faucet spout by unscrewing the ring at its base to expose the diverter valve. You can either pull the diverter valve or unscrew it from the faucet base, then take it apart for cleaning. If any of its parts are worn or damaged, replace the entire valve.

Replacing a Faucet

Installing a new faucet is not a hard chore. But you will have to work in a limited space, so it may take you longer to do than you expect. The procedure is this:

1. Shut off the water valve nearest the faucet and drain the faucet.

2. Use one wrench to grip the flexible riser tube, and put a second wrench on the fitting closest to the faucet tailpipe. Rotate the fitting counterclockwise and disconnect the faucet. The use of twin wrenches prevents twisting of the riser tube.

3. Remove the lock washers and nuts holding the faucet to the fixture. If the washers are rusted, you can clean them with any rust-removing oil.

4. Lift the faucet out of the fixture and thoroughly clean the rim of the hole in the fixture.

5. Take the lock washer off the tailpipe of your replacement faucet and apply a layer of plumber's putty to the bottom of its base. If the base unit has a gasket, no putty will be needed. Then insert the new faucet in the fixture hole and push down on it until the putty oozes out around its base.

6. Now tighten the lock washer on the tailpipe. *Be careful not to overtighten or you might crack the porcelain fixture.*

7. Connect the hot and cold riser tubes to the faucet tailpipes, using two wrenches, as when you removed them.

8. Clean all excess putty away from the base of the faucet and turn on the water supply valves. Check all your new joints for leakage. If you find any water, slowly tighten the pipe section or fitting until the leak stops.

9. Unscrew the aerator in the faucet spout and allow water to flow through the faucet to flush out any grit in the pipes or unit. Then replace the aerator.

SINKS AND LAVATORIES

"Lavatory" is the technical term for a bathroom sink. If you care about being technical you can call every other sink in your house a sink, but in the bathroom it is a lavatory. Most of the sinks in a house are made of cast-iron or steel coated with porcelain, vitreous china, plastic, or fiberglass.

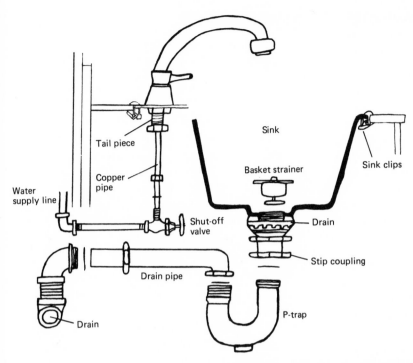

Tail piece

Copper pipe

Water supply line

Shut-off valve

Sink

Basket strainer

Sink clips

Drain

Stip coupling

Drain pipe

Drain

P-trap

Faucet hookup in a sink or lavatory.

FAUCET REPAIR CHECKLIST

Problem: Drips at spout

Possible Cause	*Solution*
Faucet not closed fully	Turn faucet handle until tight.
Defective handle or stem	Remove handle and examine knurling in handle. If it is stripped, replace the handle. Examine the threads on the stem. If worn, replace the stem.
Worn seat washer	Remove the handle and faucet stem and examine the seat washer. If it is worn, chewed, or hardened, replace it.
Worn seat	Remove the faucet stem and examine the seat inside the faucet body. If it is worn, abraded, or defective, use a seat dresser to regrind the seat until it is smooth. Many faucets have removable seats that may be replaced without disturbing the faucet body.

Problem: Spout drips, faucet noisy and vibrates

Possible Cause	*Solution*
Faulty stem or loose seat washer	Remove the faucet stem and examine it for defects. If it is faulty, take it to a plumbing supply store and purchase an identical replacement. If the seat washer is loose, tighten the brass screw. If noise and vibration persist, the threaded portion of the body may be worn and the entire faucet must be replaced.

Problem: Handle turns, water continues to run

Possible Cause	*Solution*
Defective handle or stem	Remove the handle and examine the knurling and threads on the handle and stem. If either are worn, replace the unit.

(Continued)

FAUCET REPAIR CHECKLIST (Continued)

Problem: Water leaks from handle

Possible Cause

Faulty bonnet packing or defective bonnet washer

Solution

Remove the handle and bonnet and examine the bonnet packing or bonnet washer. If the packing is compressed, replace it with new packing. If the washer is worn, replace it.

Problem: Combination faucet leaks at swivel

Possible Cause

Loose spout bonnet

Solution

Wrap the spout bonnet with friction tape, and slowly tighten it with an adjustable wrench until the leak stops.

Defective spout packing or washer

Remove the spout and spout bonnet and inspect the bonnet packing or washer. If either is defective, replace it.

Problem: Combination faucet leaks at base

Possible Cause

Faulty mixing chamber

Solution

Replace the faucet. It is almost impossible to repair a mixing chamber successfully.

Problem: Reduced water flow from spout

Possible Cause

Clogged aerator screen

Solution

Unscrew the aerator nozzle threaded to the end of the spout. Clean the aerator screen thoroughly. If the screen is rusted or broken, replace it.

Faulty nozzle

The screen and washers attending the nozzle may be so worn or coated with mineral deposits that the nozzle must be replaced. Take the nozzle with you to a plumbing supply store to be sure you purchase the proper size and type of replacement.

Kitchen sinks can be any of these materials as well as stainless steel.

No matter how they are constructed, sinks and lavatories come in all sorts of shapes, sizes, and colors; and they all have to have faucets attached to the water supply system. Just as important, they must also be connected to the house DWV system. The drain hole in the bottom of the sink is used to connect a short length of 1½-inch or 1¼-inch diameter pipe known as the tailpiece. The tailpiece fits inside one end of a P- or S-shaped pipe that forms the trap. Sometimes there is a cleanout plug threaded into the bottom of the curve of the trap; it facilitates cleaning the trap, but if there is no plug and you have a desperate desire to get inside the trap, you can easily loosen the slip couplings that hold it to the tailpiece at one end and the drainpipe

at the other. It is not mandatory, by the way, that drainpipes be assembled with sealers, or even tightened much more than you can do by hand, **since the water that goes through them is rarely under pressure or even in the pipes for very long. Bear in mind, however, that the bottom of the** curved trap is always supposed to retain some water so that vermin and gases cannot come into your house by crawling up the DWV system.

Unclogging Drains

The most common repairs to sinks and lavatories involve unclogging their drains. When water refuses to go down a sink drain, your first step is to clear away all the debris that may have collected in the sink over the drain. The next step is to use a rubber force cup (plunger).

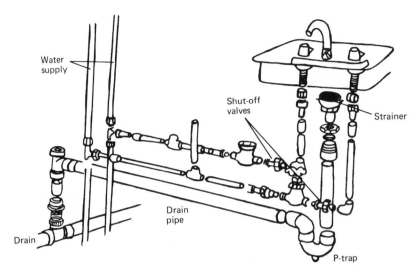

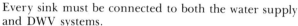

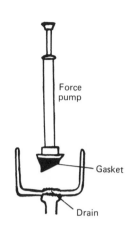

Every sink must be connected to both the water supply and DWV systems.

The compressed air cleaner.

HOW TO USE A PLUNGER

A good plunger will unclog most stoppages. The best kind has a fold-down rim so that you can use it on both sinks and toilets but they all work well. A plunger is most effective when there is several inches of hot water in the basin to act as a seal and also help to dissolve any grease that may be in the drain.

Remove the stopper in the sink. Lavatories and bathtubs have overflow openings, so plug them with a damp cloth so that you don't lose pressure from the force cup through the overflow. Tilt the force cup to one side so it can fill with water, then position it squarely over the drain. Push the handle down and up, five or six times. On the last upward stroke, yank the plunger away from the drain. You may get a geyser of water, so keep your face out of the way, but you are also exerting tremendous pressure on the clog. If you have plunged away for a while and the drain remains clogged, you still have some options:

CHEMICAL DRAIN OPENERS

Try pouring a small can of lye down the drain. Wait about half an hour and then flush the drain with water. Lye is sold at most hardware stores and is the one chemical professional plumbers will recommend. There are also a number of liquid chemicals that you could purchase, but don't. They are dangerous to the health of both pipes and people.

If you have no choice but to use one of the liquids, follow the instructions on the container precisely. If you do exactly what the instructions tell you to do, your health will not be impaired, although the drainpipe still may be eaten away.

COMPRESSED AIR CLEANER

You can also buy a compressed air cleaner, which looks like a bicycle pump without the air hose. To use it, run an inch or so of water into the sink and insert the tapered rubber end of the tool into the drain opening. Now push down the plunger. The compressed air will exert potent pressure on the clog. It also puts pressure on the drainpipes and if they are old or at all weak, don't be surprised if you blow them apart. It is *not* recommended that you use a compressed air cleaner on any system that is more than 25 years old.

THE CLEANOUT PLUG

If the drain has a cleanout plug you can unscrew it with an open-end or pipe wrench. Bearing in mind that the purpose of every trap is to hold water, don't forget to put a pan under the plug before you loosen it. With the plug removed, you can reach up into the trap with your finger, a screwdriver, or any other thin tool and clean out whatever debris is in the trap. But that may not be

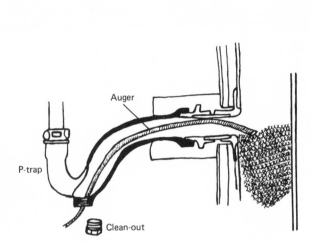

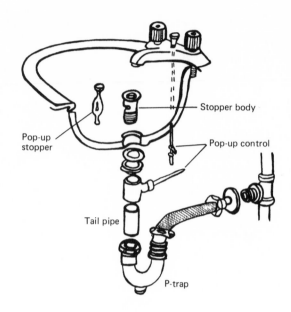

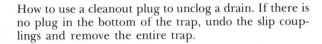

How to use a cleanout plug to unclog a drain. If there is no plug in the bottom of the trap, undo the slip couplings and remove the entire trap.

Anatomy of a sink drainage assembly.

where the clog is, at which point you will have to resort to a drain auger or snake.

AUGERS

A drain auger, or snake, is a flexible steel cable or tape that can be pushed down the drain of a sink, or through the cleanout plug at the bottom of a trap. It will follow the curve of the waste pipe and go all the way to the main stack. The tape version cannot be rotated but must be pounded against the clog. The snakes are made of cable and have a handle that lets you turn the head of the snake so you can drill into the clog.

You simply insert the end of the auger down the drain or into the cleanout plug and keep pushing until it meets resistance and then start pounding (if you have a tape) or drilling (if you are using a cable) until the clog is broken up. If, by the way, there is no cleanout plug in the trap and the trap is in your way, remove the entire curve by loosening the slip couplings that connect it to the tailpiece at one end and drainpipe at the other. Then insert the auger directly into the drainpipe.

REPLACING A SINK

Replacing a sink is neither a difficult, nor lengthy chore. It does not even require any special tools. The procedure suggested here may have to be varied slightly depending on the type of sink you are installing and the piping that attends it:

1. Turn off the hot and cold water supply.

2. Disconnect the water supply lines from the faucets.

3. Free the drain trap by loosening the slip couplings with an adjustable wrench.

4. If the sink rests on a commode, you can now lift it up, unless it has clips under its rim that hold it to the stand, in which case loosen the screws that hold the clips in place, then lift the sink off its stand. If the sink is fitted in a counter top, it will definitely be held in position by clips. There will also be caulking under the rim of the sink, which may have to be loosened by prying under the lip of the sink with a putty knife. Wall-hung sinks rest on a support bar anchored to the wall. Pull the sink upward and lift it clear of the bar. Many older

lavatories use a 1¼-inch diameter waste pipe, while all modern fixtures use 1½-inch piping. You can adapt the 1½-inch tailpipe to a 1¼-inch drainpipe by using a 1¼-inch × 1¼-inch adapter. If your new sink is different size or shape than the one it is replacing, you may also have to change the counter top, hanging brackets, or the entire commode, which means you are in for some carpentry work.

5. With the sink in position, follow the manufacturer's installation instructions for attaching the clamps. If you are installing a counter top sink, put down a bead of putty around the edge of the bowl. The trim ring around the sink will compress the putty into a watertight seal when you tighten the clamps.

6. Insert the faucets in their holes using either washers or putty and tighten the nut on the tailpipe against the underside of the sink.

7. The faucet tailpipes may need more than one connector to complete their hookup to the water supply system and some connections must be made with a sealer. The easy-to-use ⅜-inch flexible riser tube requires no sealer.

8. Apply putty between the drain and the bowl and tighten the large ring nuts beneath the drain, against the bottom of the sink.

9. Wrap the male threads of the chrome tailpiece with pipe tape or putty, or coat its fine threads with silicone rubber sealout (RTV), and screw it into the bottom of the drain. Then attach the rubber gasket, the tailpiece, and slip nut to the bottom of the fixture drain opening. The gasket resides between the tailpiece and the drain.

10. Remove the pop-up drain assembly before you install the lower drain unit. The lower drain must be sealed with tape, putty, or RTV before you screw it into the bottom of the drain. Be careful to aim the T containing the stopper valve toward the back of the fixture so that the pop-up assembly can join the drain-plunger connecting rod. When the lower drain is in place, you can connect the pop-up assembly to the pop-up rod.

11. One end of the trap is connected to the bottom of the lower drain tailpiece and can be sealed with putty or RTV placed inside the slip nut. The gasket resides between the slip nut and trap bend. The other end of the trap connects to the trap arm, which leads to the waste pipe in the wall. There it is kept leak-free by a trap adapter.

12. When you have completed all of the water supply and DWV connections, turn on the water and inspect your handiwork for leaks. If you find any, tighten the fittings with a wrench until the leak stops.

BATHTUBS

Bathtubs are manufactured from cast iron, steel, porcelain, or fiberglass; and you can buy them in practically any size, shape, and color. Generally, they are between 4 and 6 feet in length and from 12 to 16 inches deep.

Cast-iron tubs have a porcelain enamel coating baked on the metal. They are not only the most durable kind of tub, but they also weigh as much as 500 pounds; so if you are planning to install one, first make sure the floor is strong enough to support it.

Steel tubs are also porcelain-coated but they weigh only about 100 pounds, which most floors can support. The nature of steel is such that the tubs tend to be noisy, so many models are sold with a special undercoating that helps deaden the sound of any water poured into the unit.

The molded fiberglass tubs are smooth, which makes them slippery, easily cleaned, and just as easily scratched. They are also very light, and there are kits available that allow you to remove (or at least hide) the scratches.

Bathtubs are positioned under faucets attached to the water supply lines and their drains must be connected to the DWV system. The faucets for a bathtub are usually sunk in a wall above the drain end of the tub, and you will have to hunt for the water supply valves that service them. Often, the valves that control the lines servicing a tub are in the basement, directly under the tub. If the tub is located above the first floor, look for an access panel in the wall where the faucets emerge. The panel may be next to the tub or on the back side of the wall.

Bathtub faucets are likely to have a large escutcheon, or cover, that must be removed before you can do any repairs to the faucet. You can use vise-grip pliers to unscrew the escutcheon; in some cases you can pry it away from the wall with a screwdriver, once the faucet handle has been removed. With the handle and escutcheon removed,

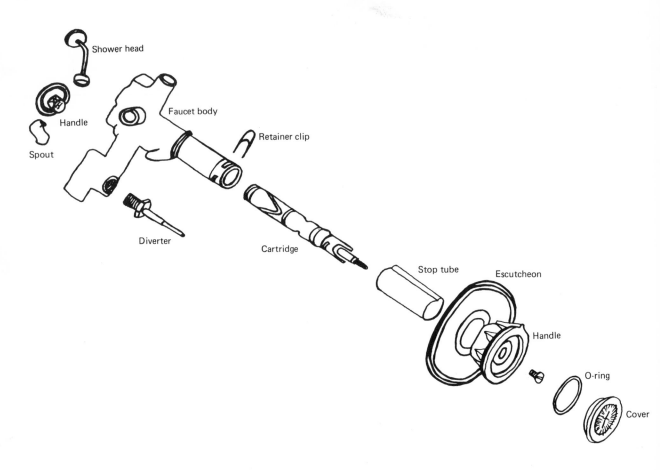

A bathtub single-handle mixer faucet will rarely develop
a leak; if it does, the entire cartridge must be replaced.

the faucet can be repaired in the same manner as
you would any sink or lavatory faucet.

Some bathtubs have a single-handle mixer
faucet consisting of a metal cartridge positioned
behind the mixing knob. If this type of faucet
develops a leak, the only repair you can make is to
replace the entire cartridge. First, take off the
handle (which may have a threaded cap over its
screw) and remove the escutcheon. Then pull the
stop tube off the faucet body. Pry the retainer clip
up and out of the front of the body. Now you can
slide the cartridge outward. When you insert a new
cartridge, push it all the way in until the front
edges of its ears are flush with the body. The
retainer clip legs have to straddle the cartridge ears
and fit into the bottom slot in the body so that the
cartridge is locked in place.

Tub Drains and Traps

When a tub drain is clogged, first try a plunger. If
that fails you have to locate the tub trap and open it
so you can get an auger into the drain and clean
out the blockage. Many tubs have a standard P- or
S-shaped trap, which may be under the floorboards and must be reached through an access
panel either in the floor or the wall, or from the
basement or crawl space. You may also find a
drum-shaped trap located under the floor next to
the tub with an access panel covering it. The cleanout plug at the top or bottom of the drum trap can
be unscrewed for entry into the drum.

Blockage in tubs often occurs at the stopper
mechanism instead of the trap; so whenever you
encounter a clogged tub drain, start by checking

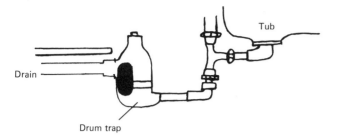

How a drum trap is connected to the tub.

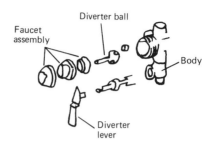

What makes a diverter work and how it comes apart.

the stopper. Most tub stoppers incorporate a spring that keeps pressure on the rod assembly that raises and lowers the stopper. The spring can collect enough debris to block the drain. In order to reach the spring, you must unscrew the plate covering the base of the stopper lever, then pull the entire lever mechanism out of its hole and thoroughly clean all of its parts before replacing it.

Many tubs have a weighted stopper that can also accumulate dirt and prevent the weight from fitting snugly into its seat. You can tell this has happened when water trickles out of the tub after the drain is closed. Remove the escutcheon and pull the stopper, and its weight out of the tub. Then clean the entire unit, especially the bottom of the weight. Before you put the stopper back in its place, run hot water down the drain for several minutes to wash away any dirt that may be on the stopper seat. While you have it in your hands, also examine the stopper mechanism for a worn or broken spring or other defective parts that should be replaced.

Diverters

Diverters are faucetlike devices used to channel water from the tub spout to a shower head and you can disassemble one in the same way you would any faucet. The things to look for when you are examining a diverter are a worn O ring, compressed packing, or a defective packing washer.

Replacing a Bathtub

If you are removing one of those old bathtubs that stands on feet you are lucky, because all its fixtures and connections are probably out in the open. But if the tub is built against two or three walls, you will have to open at least the wall behind the fixtures so you can get at the plumbing. Once the pipes are accessible, shut off the hot and cold water valves and disconnect the water supply and DWV lines. The drainpipe can be removed at the tub if you unscrew the strainer basin. With many tubs you will find notches on the inside of the strainer basin. Insert a spud wrench into the notches and rotate the basin counterclockwise. If you don't find any notches, place a large screwdriver between the bars of the drain and use it to rotate the basin. The strainer basin is normally threaded into the waste pipe, so by removing it you will free the tub from the drain. You will usually also have to disconnect the overflow pipe before you can pull the tub out of its position.

When you install a new tub, make certain it is level, both lengthwise and sideways. Tubs are supported on three sides by 1 × 4 boards nailed to the wall studs. The edges of the tub rest on the wood, but the unit must also stand solidly on the floor. If it does not touch the floor at all points, shim the bottom with pieces of wood. Don't be surprised if you have trouble getting the unit entirely level. To begin with, the floor of your house is most likely off level in at least one direction, and probably two. To make matters worse, it is the nature of large metal castings to be somewhat distorted. So if you have to have one end of the tub out of level, make sure the unit pitches toward the drainpipe so that water will always run out of it.

When you have the tub as level as you can make it, mark the top of the tub on the wall studs and also locate the center of the overflow and drain holes. Nail whatever shims you have used in so you will not have to level the tub again, and pull the unit away from the wall. The 1×4 boards should be nailed to the wall studs *below* the lines

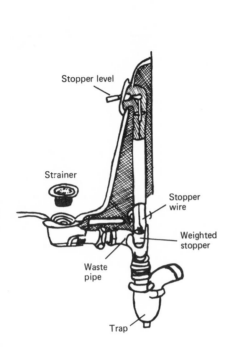

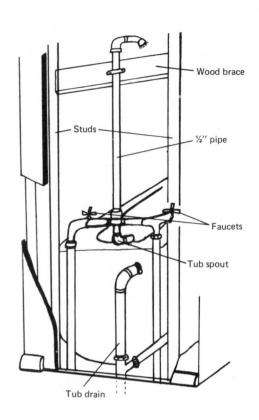

A typical tub overflow, stopper, and drain connection.

Shower connections are normally nothing more than an extension of existing plumbing.

you marked on the studs. Measure the thickness of the tub edges. That measurement is how far below the lines you place the 1×4 ledge. When the ledge is nailed to the studs, make whatever alterations are necessary to the floor.

If you have enough space to work in behind the tub, you can attach all the water supply and DWV fittings once the tub is in place. If you are tight on space, you will have to assemble your pipes and fittings in the wall so that you have as few connections as possible to make once the tub is in place. If this is the case, assemble the drainpipes from the overflow and the tub drain and also connect the trap to the waste pipe. The U portion of tub traps has a joint that allows it to rotate 180°, which gives you considerable flexibility when you are trying to align it with the tub. By measuring carefully, and with a little luck, you can connect the trap to the drainpipe and the tub drain, and line them up so that when you put the tub in place the overflow and drainpipes will be in position for their final connections. The washers that must be

against the face of the two connecting drains can be taped in place against the ends of the pipe.

While you still have room to work in, this is also the time to make your connections between the hot and cold water faucets and the water supply lines. When these are complete, put the tub in place and connect the overflow and drainpipes. Turn on the water supply valves and run water through the tub to make certain all of your connections are watertight.

SHOWERS

You can install a shower above any bathtub just by continuing the water supply lines vertically so that water can be delivered to the tub from a spout approximately 6 feet above the floor. You can also put your shower in a stall, which you can either build from plywood and cover with tiles, or you can buy one prefabricated. When you extend the water supply lines, the hot and cold lines are connected to each other by a valve diverter that channels the water away from the tub faucets and up to the shower head. You can also have independent hot and cold water pipes with separate faucets, in which case there is no need for a diverter valve.

The prefabricated metal shower stalls are sold in 30-, 32-, or 36-inch squares and consist of a stone base with a hole in its center for the drain, three sheet metal sides, a door frame, faucets, and a shower head. You can put a stall any place inside or outside your house, so long as it can be connected to the plumbing.

Shower stalls are also made from fiberglass and come as either a completely molded unit, or in separate parts that are easily assembled. Fiberglass stalls are lighter and easier to clean than the metal types, and are becoming quite popular.

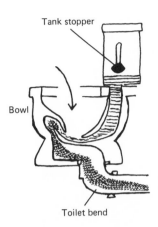

What happens to the water in a toilet when it is flushed.

Repairs to Showers

All the repairs necessary for a shower are identical to either unclogging a sink or repairing a lavatory faucet. Occasionally, a shower drain will become clogged with hair or with mineral deposits from the water and need to be cleaned. The faucets are usually standard lavatory-type faucets positioned horizontally in the wall and can be repaired following the procedures described earlier in this chapter.

Putting in a Shower Stall

The shower's floor must be level, so if your floor is not absolutely true in all directions, you will have to shim the pan. You can attach the drainpipe in the same manner as you would a sink, and then position the pan over a 2-inch diameter waste pipe with P trap rising up from the floor. Now assemble the sides of the stall around the pan, following the manufacturer's instructions that come with the unit. Place the side with the faucet holes nearest the water supply pipes and bring both lines up to the holes so that you can connect the faucets. The faucets are part of a single fitting that allows a ½-inch pipe to continue vertically up to the top of the stall where it is capped by an elbow and the shower head. When the water supply lines are connected, run water in the shower and check all your connections for leaks.

Add-on Showers

If you are adding a shower to an existing bathtub, you will most likely have to open the drain end wall behind the tub so you can extend the water supply lines. This means you may have to remove a considerable amount of tile, plaster, or sheetrock.

When the wall has been opened, disconnect the tub faucets from the supply lines and replace them with a two-valve diverter. Take both faucets out of the wall and disconnect their bodies from the supply lines. Attach the two-valve diverter to both supply lines and insert faucets in both its sockets. A ½-inch vertical fountain pipe is threaded into the center of the diverter. Bring the pipe through the wall approximately 5 feet above the tub, and attach a shower head to its open end.

An alternative arrangement is to continue each of the water supply lines vertically above the tub fixtures. Join the pipes about 30 to 36 inches above the floor, with a diverter, then carry the shower pipe to whatever height you wish. When all connections are completed, turn on the water supply valves. Run water through the installation to check for leakage in the connections.

TOILETS

Although you will find a considerable variety of design, toilets all consist of a vitreous china bowl

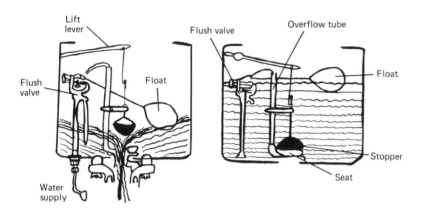

The flapper flush valve is standard equipment in most new toilet tanks and may be made of brass or plastic.

that is usually placed on the floor and connected to the DWV by a waste pipe at its base. The rim of the bowl supports a molded seat and cover and behind that is a vitreous china or plastic water tank attached to the bowl as well as to the cold water line. Inside the tank is a flush valve assembly that automatically controls any water entering or leaving the unit.

When you flush a toilet, the handle lifts a flapper or a rubber flush ball at the bottom of the tank, so that water can drain into the bowl. As water leaves the tank a float (usually a hollow ball) descends with the water level, which in turn opens an inlet valve attached to the water supply line. Once all water has drained out of the tank, the flapper or flush ball falls back in place and seals the tank. At the same time, new water fills the tank via the inlet valve. As the water level rises, the float is carried upward again until it closes the inlet valve and stops the flow of water.

Mechanical variations in toilets are usually confined to the different types of flush valves used to control the water supply and it is the valves where most toilet repairs must be made.

Flapper Flush Valve

This is the type that usually comes with the tank. It may be made of brass or plastic and consists of an inlet valve attached to the top of a water supply pipe, a bowl refill pipe (usually a small, flexible hose or tubing), a float ball, rubber flapper, and flapper lift chain or rod assembly.

Fluid Level Control Valve (FLC)

The FLC valve functions in the same manner as the flapper flush valve, but it does not have a floating ball, ball cock, or float assembly. Instead, the inlet valve is controlled by direct water pressure against a plastic float, which allows you to control how much water is used for each flush. The unit is also noise-free.

Pressure-Controlled Fill Valves

An alternative to both the traditional tank flush valve and the FLC is the new pressure-controlled fill valves. The valve itself is about 4½ inches long and 5 inches high, and in order to operate properly it must be completely submerged in the tank water. The fill valves are a product of space-age technology made of durable plastic and other noncorrosive materials, so after they are installed minerals in the water should never affect them. They function on an age-old principle: The valve remains open allowing water to enter the tank until the weight of the water on top of the valve closes it.

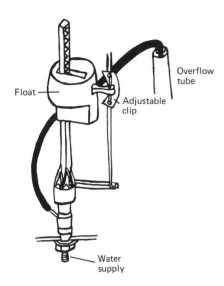

The fluid level control (FLC) valve eliminates the need for a float ball.

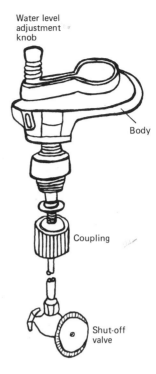

Pressure-controlled fill valves are the most compact and newest valve on the market and work just as well as their older competitors.

There is an adjustment screw on the unit that permits you to open or close the valve to control how much water enters the tank.

Like all water devices, the fill valves may experience clogging from rust, impurities, or debris in the water, which prevent it from shutting off the water as positively as it should. When this occurs, you can remove the top of the valve by undoing the pair of retaining screws and prying off the lid. There is a small rubber disk inside the cover, which you can lift out with your fingernail and wash in clean water. When you replace the disk, be certain its smooth side is showing, then replace the cover and install the set screws. Be careful not to overtighten the screws.

Replacement Tank Flappers

The flapper or flapper ball that covers the tank outlet can cause a variety of problems. They are usually made of rubber and can become corroded, chewed up, or get out of line; all result in a constant trickle of water into the toilet bowl. They can be adjusted and realigned by straightening the rods, if a rod system links the flapper to the lift handle. The chain, if that is used, can be lengthened or shortened. Or, the flapper may have swung out of line with the water outlet. Look at the flapper hinge (which is commonly some form of

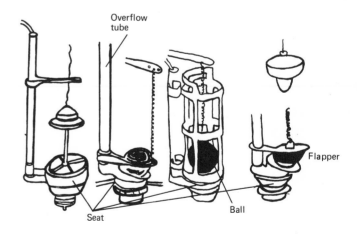

Some different types of toilet tank flappers.

ring around the bottom of the overflow pipe) to be sure it is in the proper position to allow the flapper to work correctly.

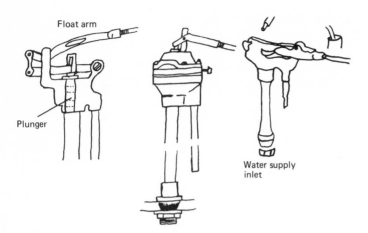

Details of some of the flush valves you may find inside
your toilet tanks.

The Problems with Replacement Flappers

Any number of replacement flappers are manufactured and purport to fit all or most tanks.
Don't bet on it. There have been some subtle
changes in the design of overflow pipes and the
way they are attached to the tank. Some of the
pipes, for example, have a high base that prevents
a new flapper being slid down the pipe so that it
will seat squarely against the bottom of the tank,
which means it will leak water. Most of the replacement flappers have a rubber ring that fits
around the overflow pipe, as well as a loop molded
on each side of the ring that can fit over the pair of
ears that protrude from the base of some overflow
pipes, for example, have a high base that prevents
a new flapper being slid down the pipe so that it
somewhat awkward to do, since making sure
means you ought to take the entire toilet tank to
the store with you, or at least you will have to
remove the tank from the bowl and take the pipe
assembly.

If you are not certain whether a new unit will
fit, look for the kind of flapper that can be fitted
over the rim of the tank drain and held in place
with epoxy solvent. These units are more expensive than just a flapper, but they can be positioned
in the drain independently of the overflow pipe
and come with the epoxy sealant. They are installed by kneading the epoxy into a ¼-inch diameter rope, which is placed around the rim of the
drain. The flapper unit is then pressed down on
top of the epoxy, which instantly seals it in place so

that the flapper is immediately ready for use. Even
with this kind of complete replacement kit, you
may have one of the four or five types of drain
seats that require a special flapper unique only to
that particular drain. If you cannot find a flapper
to fit the drain seat, your only alternative is to
remove the toilet tank and replace the overflow
pipe assembly.

Toilet Tank Repairs

It is possible to repair a faulty flush valve, and even
to purchase individual replacement parts, such as
lift wires. But replacement flush valves only cost
about $5 to $7. At that price it is simpler to put in a
whole new unit and almost as economical. There
are, however, a few minor repairs that can be
performed successfully without installing a new
unit.

Flush Valve Replacement

Replacement valves are made in brass or plastic
and both will provide long years of service. If you
need to replace a flush valve, consider buying a
fluid-level-control (FLC) or a pressure-fill valve as
your replacement. Both units are made of strong,
space-age plastic and eliminate the float ball in the
tank. Moreover, they have a very definite action
that stops the inflow of water completely. The
procedure for installing any flush valve system is
the same.

FLUSH VALVE ASSEMBLY CHECKLIST

Problem	*Solution*
Water continues flowing. Tank does not fill.	Check the handle, trip lever, lift rods or chain, flapper, and tank float ball for binding. If any part is binding, bend it so that it can move freely, or replace the part. Check the tank ball or flapper to be sure it fits into the valve seat properly. If the valve has a tank ball, it is raised by lift rods that go through a guide arm attached to the overflow tube. The guide rod can be loosened and rotated to realign the lift rods. If this fails, purchase a rubber flapper, which you can slide down over the refill tube and replace the ball. The flapper has a chain attached to the end of the trip lever.
Tank fills. Water continues to flow.	Lift the ball float. If that stops the flow (it is probably making a hissing sound), bend the float arm downward slightly. If this fails, remove the two set screws in the inlet valve assembly and pull the inlet valve out of its seat. Check the valve seat for corrosion, which can be cleaned with fine sandpaper. Replace the small washer on the valve if it is worn or deteriorated. Also check the tank ball or flapper to be sure it is seating properly.
Water level too high or too low.	The water should be about one inch below the top of the overflow pipe. Bend the float arm up to raise the water level and down to lower it. If you have an FLC valve, the float is regulated by moving a small clip up and down on its traveler rod.
Partial flush.	The tank flapper or ball is not opening enough. Bend the upper lift rod or shorten the lift chain.

Close the water supply valve, which is usually positioned directly under the tank. Flush the toilet to empty the tank and sponge out any remaining water. Unscrew the nut under the tank that attaches the tailpipe of the existing flush valve to the water supply pipe, then undo the nut under the bottom of the tank that holds the flush valve tailpipe to the tank. You can now lift the flush valve and ball float assemblies out of the tank. There is a rubber refill tube clipped to the top of the overflow pipe. Simply pull the clip off the pipe. The new flush valve comes with a rubber washer that slides over the valve tailpipe and fits against the bottom of the valve. Put the washer in place and stand the valve in its hole in the tank. You will have to hold the unit upright until you have hand-tightened the nut on the tailpipe under the tank. When the valve is in place, tighten the lock nut a half-turn further with a pipe wrench. The flexible riser tube that leads from the water supply valve may have a flared end, or it may be straight. Most replacement valves come with the appropriate washers and nuts for assembling either to the valve tailpipe. Essentially, you must place a washer over the end of the tube, which is then united to the tailpipe with a reducing nut. Use a wrench to tighten the nut one half-turn further than you can tighten it by hand. If the new unit is an FLC or pressure-fill valve, it is ready to work as soon as you push the refill tube over the nipple at the top of the valve assembly and clipped its free end to the top of the overflow tube. If the new unit is a float valve, you can unscrew the ball float and float rod from the old assembly and attach them to the new one. Replace the tank top and turn on the water supply valve. Check all your connections as the tank is filling to be sure they are

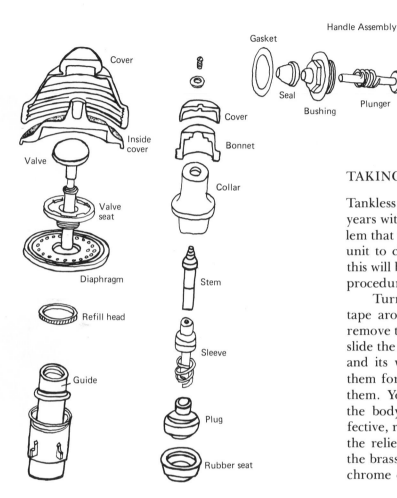

Anatomy of a tankless flush valve.

TAKING APART A TANKLESS FLUSH VALVE

Tankless flush valves are constructed to last for years without needing any repairs. The only problem that will arise is water leakage or failure of the unit to close automatically, and in most instances this will be the result of worn-out parts. Follow this procedure for disassembling a tankless flush valve:

Turn off the water supply valve. Wrap friction tape around the chrome valve cover before you remove the cover with a pipe wrench. You can now slide the inside brass cover, the rubber diaphragm, and its washer off the center post and examine them for wear. If any parts are defective, replace them. You can also pull the relief valve out of the body. Examine it for wear and, if it is defective, replace it. In reassembling the valve, insert the relief valve, then the diaphragm, and finally the brass inside cover on the body, then screw the chrome cover back in place.

not leaking. If any of them leak, tighten the nuts until it stops. Adjust the tank water level.

Tankless Flush Valve Toilets

Tankless flush valve toilets have no flush tanks. The flush valve is designed to release water directly into the toilet bowl, and after a few moments to shut itself off. There are two types of flush valves, one using a diaphragm and the other employing a piston. The replacement parts for both types can be purchased in kit form, which allows you to rebuild a worn-out valve. If the entrance line bringing water into your house is more than 1¼ inch in diameter, it is possible for you to install a tankless flush valve toilet. Otherwise, your plumbing system is not large enough to handle the unit.

DISASSEMBLING THE FLUSH VALVE HANDLE

Wrap friction tape around the chrome nut at the base of the handle and loosen it with a pipe wrench so you can remove the handle. Be careful to note the exact position of all parts inside the handle base so that you can reassemble them properly. Take off the ring-type face washer. If it is worn, replace it. Unscrew the inside nut that holds the plunger and handle together and remove the bushing that retains the handle spring and seal. Replace the spring or seal if either is worn. Start putting the handle together again by securing the spring and seal to the bushing. Insert the plunger in the bushing and lock the handle to it with the handle nut. Position the ring washer and insert the handle assembly in the valve body. Tighten the chrome collar around the handle.

TANKLESS FLUSH VALVE CHECKLIST

Problem: Drips at handle

Possible Cause	*Solution*
Loose collar nut at handle base	Wrap the chrome nut with tape and tighten the collar nut slightly. Be careful not to overtighten.
Ring washer in handle assembly defective	Disassemble the handle. If the ring washer is worn, replace it.
Handle spring or plunger defective	Disassemble handle. If the spring or plunger is worn or broken, replace it.

Problem: Water continues running

Possible Cause	*Solution*
Diaphragm dirty or damaged	Disassemble valve unit. Wipe the diaphragm thoroughly. If there are tears or rough spots in the rubber, replace the diaphragm.

Problem: Valve does not flush

Possible Cause	*Solution*
Relief valve defective	Disassemble valve and examine the interior relief valve for wear or damage. Replace if faulty.
Handle assembly defective	Remove the handle assembly. Examine all parts for wear or defects. Replace damaged parts.

Removing a Toilet

In the event that you are replacing an old toilet, the connections are not difficult to make, providing the new toilet has the same rough-in as the old one. Rough-in is the distance from the finished wall to the center of the flange that connects the toilet bowl to its drain. In most cases the rough-in is 12 inches. You can determine the rough-in of any toilet by measuring from the face of the finished wall to the nearest bolt in the base of the unit.

When dismantling a toilet, first shut off the water supply valve. Flush the toilet and sponge out both the tank and bowl. Next, undo the coupling that connects the water supply to the water tank. If the toilet is an old one, the tank is probably linked to the bowl with an ell and has some sort of connection on the wall to support the tank. First loosen the ell couplings, then all of the connections holding the tank to the wall. You can now lift the tank away from the bowl. There are porcelain caps glued to each side of the toilet bowl base. Pry each of them up with a screwdriver blade and loosen the nuts beneath them.

You may need a chisel or knife to scrape the sealing compound around the base of the bowl so you can loosen it from the floor. When it is loose, the bowl can be lifted vertically off the bolts. The bowl is heavy, so get a good grip on it. Now remove the bolts and gasket that surround the lead drain-pipe. The bolts will come free if you rotate the ferrule until their heads are centered in the wide part of the slots. Clean the ferrule and the area around it, and examine the lead bend carefully for any signs of damage. If the lead is split, scrape the area with a knife until the lead is a bright silvery color. Rub ordinary shoe polish (any color will do) around the brightened area so that you create a kind of dam to keep the solder from running off the pipe. Use a propane torch to heat the damaged area, then hold a bar of solder near the damaged spot. Apply flame to the solder until it begins to sag, then rub it against the lead bend and leave a

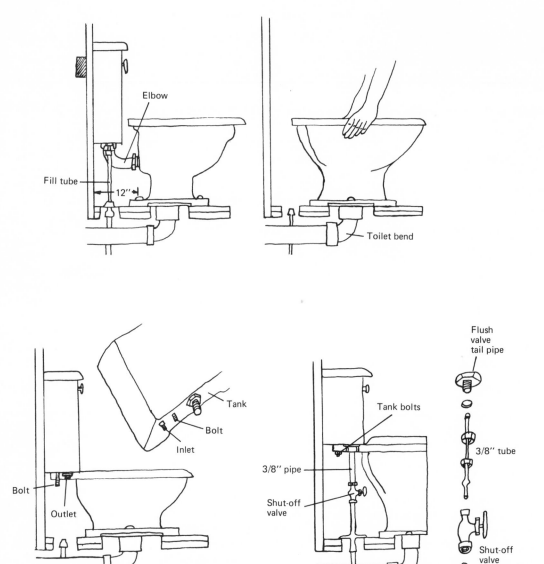

Procedure for replacing a toilet.

mound of solder over the damaged area. Fold a piece of paper into a pad and wipe the solder smooth. If necessary, reheat the solder and continue wiping until you have a smooth patch.

When the ferrule plate is absolutely clean, you can replace it on the bolts. Rotate the plate until the bolts are seated in the narrow portion of the slots. The bolts should be vertical and in line with the holes in the base of the toilet bowl. If they do not stand upright, you can hold them in place with globs of putty around their bases.

Insert a wax toilet gasket over the outlet horn of the toilet bowl. Position the bowl squarely over the flange and its bolts. Hand-tighten the bolt nuts. When you complete your tightening of the ferrule bolt nuts with a wrench, tighten them alternately;

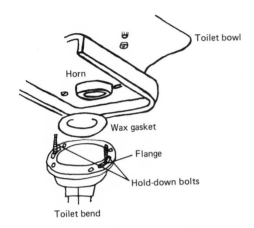

How the closet bend connects to the bottom of a toilet bowl.

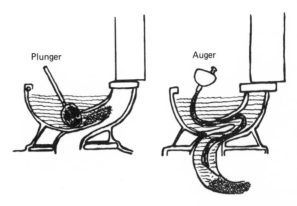

A toilet bowl can often be unclogged with a plunger. If that fails, you will have to run an auger down the toilet drain. Any auger should do the trick; a special toilet auger has a longer handle on it.

do not overtighten, or the bowl may chip or crack.

Place the tank washer around the bowl inlet opening and position the tank. Most new toilets have bolts that extend down from the tank and through holes at the back of the bowl. Tighten the nuts to the undersides of the bolts and connect the water supply line to the flush valve in the tank. Place the cover on the tank and open the water supply valve. You should flush the toilet several times to check all your connections for leakage. If any of them leak, tighten the nuts gradually until the leak stops.

Replacing a Toilet Seat

About the only repair ever needed on a toilet, other than unclogging its drain, is to replace the seat. To remove a seat and its cover, first close them, then loosen the nuts that extend from the back of the seat hinge down through the top of the bowl. The bolts are locked in place with nuts and

you will need an adjustable wrench to loosen them. When the nuts are removed you can lift the seat and its cover off the bowl.

Before installing the replacement seat, clean the bolt holes and the surrounding area. Place a washer over each bolt hole and insert the bolts on the new seat into the holes. Slide a rubber washer, lock nut, and nut over each bolt as it protrudes under the bowl and tighten them. The nuts should be tightened alternately until the seat assembly is secure, but do not overtighten or you may chip or break the toilet bowl.

UPFLUSH FIXTURES

Any sink or toilet installed in your basement that stands at a level lower than the sewer line must have some method of expelling waste upward before it can be drained out of the house. As a result, upflush sinks or toilets have special pumping equipment as part of their design.

TOILET REPAIR CHECKLIST

Problem: Water runs continuously

Possible Cause

Tank stopper does not seat or is damaged.

Ball float faulty

Faulty inlet valve

Faulty overflow tube

Solution

Observe the action of the tank stopper ball or flapper. If the stopper is connected by rods, adjust the upper rod. Check the chain to be sure it is the proper length to allow the flapper to close without getting caught under it. If the ball or flapper is worn or damaged, replace it.

Unscrew the ball from its float rod. If it is encrusted with scale or has any cracks that allow water to enter it, replace the ball.

Remove the screws holding the rocker arm on the flush valve. Examine the valve washer and replace if worn.

Unscrew the overflow tube and examine it for corrosion or leaks. Replace if damaged.

Problem: Tank will not flush

Possible Cause

Faulty lever assembly

Faulty inlet valve

Solution

Flush the toilet and observe the action of the handle, lift rod, and guide rods or chain. They should move smoothly and not bind. If the lever cannot be bent into place and if the handle is defective in any way, replace them.

Remove the inlet valve and examine it for worn or faulty parts. Either replace parts or the entire flush valve.

Problem: Toilet tank leaks

Possible Cause

Damaged tank

Faulty inlet supply valve

Faulty outlet washer or worn spud washer

Solution

Examine the inside and outside of the tank for cracks or other damage. If damaged, replace the tank.

Examine the underside of the tank where the water supply line is connected to the flush valve for any leaks at the connections or the pipes themselves. Tighten the leaking connections and replace broken pipes or replace the entire assembly.

Examine the outlet pipe area for leaks. Wipe the area dry and flush the toilet and examine again for leakage. If there is leakage, disconnect the tank and bowl and replace washer.

Problem: Toilet leaks at base

Possible Cause

Bowl is damaged

Bowl is loose

Faulty grout at base of bowl; loose ferrule

Solution

Examine surface of the bowl for cracks or large chips. If damaged, replace bowl.

Rock the bowl. If it is loose, pry off the bolt caps and tighten the nuts underneath.

Observe whether water appears around the base of the bowl. Remove the bowl and examine the ferrule plate. It should be firmly in place and level. If it is not tight, replace the ferrule plate.

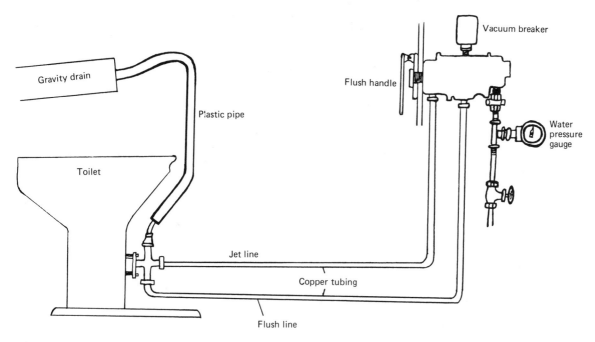

There are many versions of the upflush toilet to be found on the market.

The sinks incorporate a centrifugal pump to discharge waste water. The pump is connected to a capacitor start motor and has two valves. The drain line from the sink is connected to the pump inlet valve and an outlet pipe runs from the pump up to the sewer line. Every time you want to drain the sink you must run the pump until it has removed all waste water.

The repairs that arise with upflush sinks are most likely to be in the pump and its motor, rather than the plumbing. Faucets may develop a drip and can be fixed the same way any faucet is repaired. If the drain system clogs, it is easier to disconnect the lines and work on them when they are free, than to try and run an auger through them.

Upflush Toilets

The upflush toilets available on the market offer so many different designs that each manufacturer provides complete instructions for connecting the particular unit you purchase. Essentially, water en-

ters the upflush bowl through a two-way, or double action, flush valve connected to a ½-inch or larger cold water pipe. When the toilet is flushed, the valve first sends water through copper tubing connected to a disintegrating jet located in a T fitting at the base of the toilet bowl. The jet is powerful enough to break up all solids in the toilet and after a few seconds, the flush valve switches to its flush mode by closing off the jet and simultaneously opening the outlet to a second copper tube. The second tube delivers water to the bowl in a flushing stream that carries the liquefied waste out of the bowl, up a copper or plastic pipe, and into a gravity drain. The water pressure needed is 40 psi for a 10-foot lift, which is higher than most upflush toilets are required to perform. The gravity drain is positioned high enough above the sewer line so that it can slope $1/16$ inch for every foot of its run to the house sewer.

Complete repair and maintenance information comes with every model of upflush toilet and is found in the user's manual written for each model.

4 Pipes, Fittings, and How to Work with Them

While the fixtures in your home will demand most of your attention so far as repairs are concerned, the vast majority of the plumbing system of which they are a part is made up of pipes. Pipes carry water into and out of the fixtures. They hang there, year after year, buried between the studs of your walls or suspended from the joists in your basement. Sometimes, if the weather is humid enough, the cold water line will perspire and drip condensation on the basement floor. Once in a while a pipe somewhere develops a loose hanger and squeaks when the faucets are shut off. But by and large it is possible to spend a lifetime in one house and never even be aware that there are any pipes at all around you.

The pipes are actually very special. They are engineered to last for years. They are expected to be unobtrusive, even though the water supply lines must be under a constant pressure that amounts to between 40 and 100 pounds per square inch. Multiply all the square inches in your home plumbing system and you are talking about thousands, perhaps hundreds of thousands of pounds of pressure. The DWV pipes are not required to sustain very much pressure, but the pipes used in the DWV must meet strict standards that permit them to withstand years of corrosion. Each task required of the pipes in any plumbing system must be fully met, so over the years several metals, and recently several types of plastic, have been agreed upon as

the best, most durable ones to use. Each of these materials has a specific place in the plumbing system, and each has a place where it should not be used. Each kind of pipe also requires different techniques for cutting, fitting, and assembling into a single, watertight unit that a plumbing system must be.

FITTINGS

Fittings are those small odd-shaped pieces of pipe used to link pipes together. There is a full range of fittings available for each type of pipe used in plumbing. Given the right fitting, you can go around corners, branch off in different directions, reduce the size of the pipe run and even make chemically safe connections between different kinds of pipe. There are two categories of fittings, *standard* and *flush-walled* (or drainage-type). Standard fittings are used for the water supply lines and in the venting portion of the DWV system. When the fitting is connected to a pipe, a shoulder is created by the thickness of the pipe inserted into the fitting. The shoulders do not disturb clean water flowing through the pipes, but they are large enough to trap waste in a drainage pipe. Consequently flush-walled fittings are used in the drain-waste portion of the DWV system, because they provide smooth, unobstructed flow.

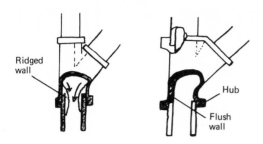

Standard fittings produce a ridge inside their connection that will trap waste and cause clogging (left), so all DWV connections are made with pipe and fittings designed to produce a flush wall at their connections.

Fittings are so numerous and varied in design that the best way to find the one you need for any given situation is go to your local plumbing supply store and describe what you want to do. The chances are the store will have exactly the kind of fitting you need and the salesperson will be delighted to sell it to you. The most common fittings used in plumbing are:

Branch Fittings

Branch fittings are called either T fittings or Y fittings because of their shape. They are used to join two pipes going in the same direction, with a third pipe that goes off at either a 45° (T fitting) or a 90° (Y fitting) angle. The third pipe does not have to be the same diameter as the other two. When buying a T or Y, always state the "run" or "through" size first, and then the branch size. For example, a Y fitting that permits a ½-inch pipe to branch off a ¾-inch pipe run would be described as a ¾ × ¾ × ½ Y.

Elbows

Elbows are used to join pipe at either 45° or 90° angles. However, when an elbow is used in the DWV system it cannot have any shoulders, and is therefore sometimes called a bend.

Reducing Couplings

Reducing couplings accept a pipe of one diameter in one end, and a smaller diameter pipe at the other. You can also buy a reducing T or Y coupling.

Bushings

Bushings allow you to connect different pipes and fittings of different sizes. A bushing fits into the larger fitting socket to accept the smaller pipe.

Crosses

Crosses are just that; they permit four pipes to come together from opposite directions, but these are rarely used in residential plumbing.

Dielectric Couplings

Dielectric couplings connect two different pipes of different metals so that there will be no undesirable electro-chemical reaction. If permitted, this would eat the joint away, eventually causing a leak.

Unions

Unions are used to connect any pipe that you expect to take apart at some future time.

Plugs

Plugs are placed into the socket of a fitting to seal it.

Caps

Caps are placed over the end of a pipe to seal it.

Adapters

Adapters are used to join two pipes or fittings of different types, such as sweat-soldered copper and solvent-welded plastic.

Nipples

Nipples come in various diameters and lengths of threaded pipe ranging from about 1 inch to 12 inches. Essentially, a nipple is a short piece of threaded or unthreaded (make it yourself) pipe, depending on the pipe you are buying.

Couplings

Couplings join two pieces of pipe and are commonly used when one length of pipe is not long enough.

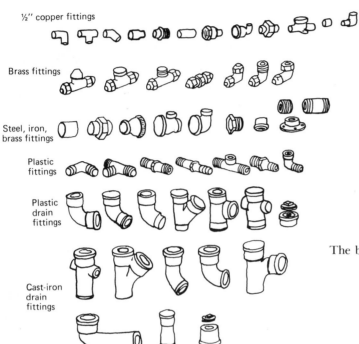

½" copper fittings

Brass fittings

Steel, iron, brass fittings

Plastic fittings

Plastic drain fittings

Cast-iron drain fittings

The basic fittings used with different kinds of pipe.

Slip Couplings

Slip couplings, also called repair couplings, are used in DWV systems to add a new fitting to an existing pipe. A *sisson coupling* serves the same purpose in cast-iron DWV pipe.

Wing Elbows

Wing elbows are specialized fittings used to bring shower arms or faucets out from the wall.

Cleanouts

Cleanouts are fittings with removable plugs that give easy, quick access into a line of DWV pipe for removal of a stoppage.

Street Fittings

Street fittings are used in both water supply and DWV; they have one end that is sized to enter a fitting socket without a short nipple placed between.

TYPES OF PIPE AND HOW TO MEASURE THEM

You will have a hard time purchasing any galvanized steel, brass, copper, or plastic pipe unless you know how to measure their inside and outside diameters, as well as their lengths. Pipe and tubing is usually referred to in plumbing supply outlets by the nominal diameter of the inside. However, a ¾-inch pipe, which you might assume to be ¾ inch across its inside, may actually be slightly larger, while its outside diameter will be about an inch. When you are determining what size pipe you need, measure the inside diameter very carefully and don't be upset if it comes out to be more than the ½, ¾, or 1 inch you expect it to be.

There are some simple tricks to measuring pipe that can make your life a little easier. Anytime you are measuring the inside or outside diameter of a pipe, use a rigid rule such as a carpenter's wooden folding rule. To measure the outside diameter, hold your thumb against the outside of the pipe and press the end of the rule against your thumb. Swing the rule back and forth in an arc

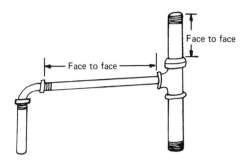

The make-up distance is the amount of pipe that is inserted into a fitting.

over the pipe and take the maximum measurement of the far side of the pipe.

If you are measuring the inside diameter, press the edge of the rule against the inside of the pipe hole and swing the rule in an arc. Take the maximum measurement of the inside edge of the pipe hole. The measurement you get is the nominal dimension of the pipe and is the one you use when buying the pipe. It will be plus or minus ½ inch, ⅜ inch, ¾ inch, 1 inch, or whatever; it may vary by as much as ¼ inch.

To measure the circumference of a pipe, you have to use a flexible rule, such as a metal tape that can be bent around the outside of the pipe. Wrap the tape *squarely* around the pipe. If the tape is wrapped at an angle, your measurement will be too long. With most steel tapes there is a hook riveted to the free end that will get in your way, so begin your measurement from some calibration other than zero. The 2-inch mark is as good as any. Wrap the tape tightly around the pipe and read the measurement that lines up with the 2-inch calibration. Then subtract the 2 inches. For example, if you have measured from the 2-inch position and the reading is 8⅜ inches, subtract 2 inches. Thus, the circumference of the pipe is actually 6⅜ inches.

The steel tape rule is also the best instrument for measuring the length of a pipe. The hook end of the tape can be attached to one end of the pipe and the tape stretched out to the far end. But before you cut any pipe you should not only measure the distance of the pipe run very carefully, but you should then measure the lengths of pipe you will need, and allow a *make-up distance* for the

fittings. The *make-up distance* is the amount of pipe threaded or inserted into the fitting, and it varies with every pipe size. When you are measuring two pipes to be connected by, say, a coupling, subtract the fitting make-up allowance from the length of each pipe when you cut it. The most accurate way of figuring the exact lengths of pipe to be cut is to lay them all out along with their fittings and then measure them again. Here are the usual make-up allowances for threaded pipe, although these can vary slightly:

> ½-inch pipe: ½ inch
> ¾-inch pipe: ½ inch
> 1-inch pipe: $9/16$ inch
> 1½-inch pipe: ⅝ inch
> 2-inch pipe: $11/16$ inch

PIPE, AND HOW TO WORK WITH IT

Plumbing systems incorporate rigid galvanized steel, brass, plastic, vitreous clay, and cast-iron pipe, as well as plastic and copper tubing. Each of these has its advantages and shortcomings, and therefore each has a specific purpose in the system where it is used.

Galvanized Steel Pipe

Nearly every old plumbing system in America is made up of galvanized steel piping. In fact, galvanized is so widely used that most modern fixtures and appliances are designed to be connected to it without adapters. Large diameter galvanized pipe is even used to make up DWV systems, although galvanized pipe should never be buried under a building or embedded in concrete where it cannot be easily replaced.

The great advantage to galvanized pipe is that it can withstand tremendous pressures, will not puncture, and is less expensive than any of the other metal pipes. On the other hand, galvanized pipe corrodes easily and may become clogged from the scale left by hard water, so many local plumbing codes now prohibit its use in any new home plumbing system.

Galvanized steel, like brass, also presents a problem anytime you need to tap a new line into the existing system. First, you have to cut the pipe twice to provide enough room for the fitting. Then

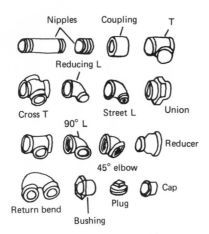

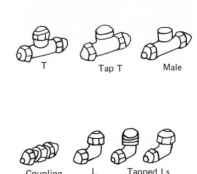

Fittings used for galvanized steel pipe are also made from galvanized steel.

Brass piping can be assembled using flared fittings such as these, as well as threaded fittings that are identical to those used with galvanized steel.

both ends of the pipe must be threaded. Next, a union must be screwed to one end so that the fitting can be threaded on. It is no picnic working in tight quarters on a galvanized steel pipe that you cannot take out of the system so you can cut and thread it. You do have one alternative and that is a saddle T fitting. You can put the saddle T over the pipe and drill a hole into the metal for your branch line. But you will get a reduced flow of water.

An easier way is to use a dresser or compression T. Like a dresser coupling, this slips over the unthreaded cutoff pipe ends and tightens leak-free. Flow is not reduced appreciably and it offers a threaded branch tapping where the new run of pipe starts. Dresser Ts are available in various sizes with full-size and reduced-size branch tappings.

You can purchase various short lengths of galvanized pipe at plumbing supply stores. These are threaded at both ends. Galvanized steel pipe, however, normally comes in 21-foot lengths, with nominal inside diameters ranging from ⅜ inch, ½ inch, ¾ inch, 1 inch and larger.

Brass Pipe

In locales where the water is so hard it corrodes galvanized steel, the local building code may insist you use brass pipe. Brass is much more expensive than other metal pipe, but it guarantees a durable water supply system. It is almost as resistant to pressure as galvanized steel and errant nails driven through a wall will not go into it. The inside of a

brass pipe is also very smooth, so that any water flowing through it meets almost no resistance. However, if you connect brass to steel pipe, you must use a dielectric coupling or union. Otherwise corrosion will quickly set in at the joint. Copper or plastic extensions added to a brass system will cause no such problems, so these can be connected without using any special fittings.

Brass pipes are also threaded. However, the act of threading brass does not remove any of its protective coating, as happens with galvanized steel. You do have to be careful when working with brass so that you don't mar its finish. To avoid any nicks or burrs from the jaws of your pipe wrench, professionals use a strap wrench on brass pipe and a regular wrench on the fittings. Three-eighth-inch brass is used with lavatories and toilets. Half-inch brass is connected to kitchen sinks and water heaters, while the ¾-inch diameter is used for main line branch lines and branch risers.

Black Iron Pipe

Black iron pipe is primarily used for gas lines. Galvanized steel pipe tends to flake from chemicals in natural gas, possibly causing the galvanized particles of the metal to clog gas valves or the tiny burner orifices in gas-burning appliances. Black iron pipes come in the same diameters as galvanized pipe; their fittings are interchangeable, but because of the flaking, galvanized fittings should not be used in any gas line. Gas mains are normally 1-inch pipe with the branch lines to hot water heaters and ranges, ½-inch pipe; and to furnaces ¾-inch pipe.

However, the pipe size you use in a gas system actually depends on the British Thermal Unit (Btu) or caloric rating of whatever appliances the pipe is serving, together with the distance of the pipe run. To determine exactly what pipe sizes to use, it is best to ask your local gas company representative for an evaluation. You should also ask for a recommendation if you are considering any unusually long-run additions to your existing system, such as you might need to install a gas lamp or an outdoor gas-fired barbecue grill.

Working with Threaded Pipe

If you are putting in a lot of piping, it might pay to invest in some pipe cutting and threading tools. While you can cut galvanized steel, black iron, and brass pipe with a hacksaw, a pipe-cutting tool is a lot more efficient, and a stock and die set is mandatory. The procedures for cutting and threading pipe with these tools are not difficult to follow.

Cutting Pipe

1. Clamp the pipe in a pipe vise. If you don't have a pipe vise, you can use a regular metal-working vise, but be careful not to squeeze the pipe out of shape.

2. Place the pipe cutter around the pipe with its cutting wheel over your cut mark. Tighten the cutter until it is snugly against the face of the pipe. Rotate the cutter around the pipe until the wheel turns smoothly and easily.

3. Tighten the cutter again. Then rotate the tool until it turns freely once more. Keep on tightening and rotating the cutter while applying a few drops of oil on the cut each time you stop to tighten the cutter until you sever the pipe.

4. Both the inner and outer surfaces of the cut end will be rough. Insert a tapered pipe reamer into the pipe, and turn it until the inside burrs have been removed. The outside edges can be smoothed with a metal file and steel wool.

Threading Pipe

1. Place the pipe end about one foot from the vise.

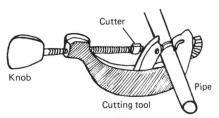

Pipe can be cut with a hacksaw, but a pipe or tubing cutter will do a neater, quicker job.

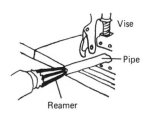

Clean the inside burrs from the end of cut pipe with a pipe reamer; the outside edge can be filed or smoothed with steel wool.

2. Put the proper size die in the die stock and lock it in place. Now push the stock over the pipe end. You will have to turn it slightly until you feel the die take hold of the pipe.

3. Rotate the die clockwise. If the die binds, back it up one quarter turn to clear any metal filings from the threads. Each time you complete a full rotation of the die, squirt cutting oil on the pipe through the openings in the die. If you cut without using cutting oil, you are likely to break the pipe threads or blunt the die.

4. When you can see approximately $\frac{1}{16}$ inch of the end of the pipe sticking through the stock face, you are finished. Rotate the stock slowly and evenly counterclockwise, until the die comes free.

5. Wipe the threads with a cloth to remove any metal particles.

Assembling Fittings on Threaded Pipe

It is easier to put any fitting on a pipe when the pipe is held in a vise. But there are many times when using a vise is impossible, so your alternative is to put one pipe wrench on the pipe, and another on the fitting. The two wrenches should face in opposite directions, as you will see when you try it.

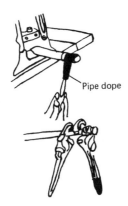

Pipe dope

If you cannot use a vise to assemble a fitting, a pair of pipe wrenches will work almost as well.

Wrap the end of the pipe with Teflon tape or smear the threads with pipe dope, and hand-tighten the fitting to the pipe. When you cannot tighten the fitting any more by hand, hold the pipe with one wrench (or in a vise) and tighten the fitting with a second wrench until you can no longer turn it without exerting extreme force. Then stop. Pipe threads are tapered. If you turn a fitting too far you can split the metal. If you count two or three exposed threads behind the fitting, you can assume you have tightened it enough.

Whether you are assembling or disassembling a fitting, either the fitting or the pipe can be turned; which one you rotate is a matter of which is easier to turn. If you are removing a fitting and normal force does not move it, apply a few drops of penetrating oil. If that fails, heat the fitting with a propane torch or as a last resort, cut it off with a hacksaw or pipe cutter.

Copper Pipe and Tubing

Copper pipe is easily installed, resists corrosion, and can be used in both the water supply and DWV systems. It is lighter than galvanized steel, and can be joined with compression or flare fittings or by soldering.

Copper is soft enough so that nails can be driven into it and it must be supported every three feet or so along its run. It is more expensive than galvanized steel or plastic piping, but offers almost as durable and reliable a system as brass.

There are two kinds of copper pipe: soft tempered and hard tempered. Soft-tempered (tubing) is flexible enough to bend around corners. Hard-tempered is rigid and therefore excellent for long runs and a neat-looking installation. The flexible tubing is normally used in short hidden runs from the water supply main to an appliance and is designated by its nominal inside diameter. The pipe and tubing sizes usually found in homes are nominally ⅜ inch, ½ inch, and ¾ inch. The outside diameter of copper pipe tubing is generally ⅛ inch larger than the inside. The inside diameters of copper tubing are usually pretty close to their nominal size although the wall thickness can vary considerably.

No matter what its inside diameter, you have three sizes of pipe and tubing to choose from. Type K is thick-walled. Type L is the wall thickness usually found in house systems. Type M is the thinnest of all and therefore the least expensive. All three types have the same outside diameter and will accept the same fittings. Type M should not be buried in the ground.

When ordering either copper pipe or tubing, you may find some confusion over exactly what size you need. Be as explicit as possible and state the length, the nominal size, the temper, and type. Types K, L, and M tubing are available in hard-tempered lengths of 20 feet. You can also buy types K and L as soft-tempered tubing, which is sold in coils of either 30 or 60 feet. Type L pipe can withstand up to 600 pounds per square inch. But the sweat-soldered joints that hold copper pipes together can only withstand 100 psi, which means that water hammer can easily blow them apart. It is essential that air chambers be located near all fixtures attached to a copper water supply system.

Copper pipe used in the DWV system is thinner-walled than the version used for the water supply system because it does not have to withstand much pressure. The diameter of copper drainage pipe is always ⅛ inch more than the pipe's nominal size, and is joined by sweat-soldering shoulderless DWV fittings to the pipe. Copper DWV pipe is available in 20-foot lengths with 1½-inch, 2-inch, and 3-inch diameters.

When assembling a copper DWV system, sweat-soldering is the only way you can join the pipes. If you are working on a water supply system using flexible tubing, you have a choice between sweat-soldering or using flare and/or compression

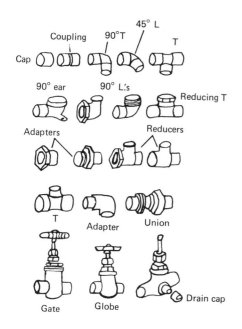

Sweat-fittings for copper piping.

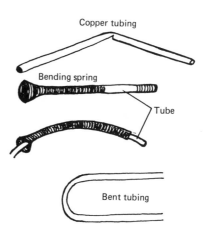

A spring-type bender can be used to curve copper tubing by inserting the tube into the spring and then bending it.

fittings. The flare fittings are easier to handle and take much greater pressures, but they are more expensive than soldered joints. Rigid copper pipe gets soldered fittings.

When you measure copper pipe or tubing, generally you need only add the diameter of the pipe to each end of each pipe to account for the make-up distance for the fittings. Thus, a ½-inch pipe running 12 feet, plus a fitting at each end would be cut 12 feet and 1 inch long. Even so, you should measure each fitting before cutting the pipe; manufacturers use various socket depths in their fittings. When you cut a copper tube or pipe, the cut must be square, as well as free from any burrs or rough edges. Since copper is a soft metal, you can cut it with a hacksaw, but you will get a neater cut if you use a pipe or tubing cutter. The burrs on the outside of the pipe can be removed with steel wool. Most tubing cutters have a triangular reamer attached to them for smoothing out any inside burrs.

Copper pipe is too rigid to be bent, but flexible tubing is a different matter. You can make long curves in copper tubing just by bending it over your knee. But if you require a short, tight bend, use a bending hickey or a special spring, which is placed around the tubing and used to

produce a smooth, tight curve. It is permissible, by the way, to bend a length of tubing between two rigid copper pipes that are too far out of line to be directly joined by a fitting. In fact, it may be a fine idea.

Sweat-Soldered Joints

Sweat-soldering is the most common method for joining copper pipe or tubing in a water supply system. The procedure requires a propane torch, which is easy to use once you have given yourself a little practice. If you intend to do some sweat-soldering, make two or three practice joints before you do anything important. Be very careful, though. A carelessly directed flame can start a fire, which can be especially hazardous since the water is cut off.

Propane torches do not deliver enough heat to heat the larger diameter pipes used in a DWV system, so for that you will have to use two of them together or else use a blow torch. However, the heat from a blow torch is so intense that it can distort the mechanism inside a valve; therefore, as a precaution always remove the insides of any valve you are soldering. Then solder the valve body in position and wait for it to cool before replacing the

Coat both the end of the pipe and the inside of the fitting with a nonacid paste flux. Then assemble the pieces.

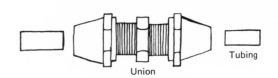

Disassemble the flare nuts from the union fitting and slide them over the ends of the pipe to be joined.

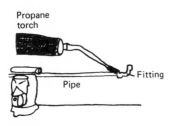

Heat the union with a propane torch until the pieces are hot enough to melt the solder.

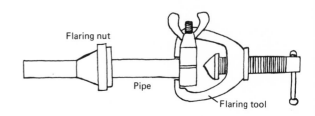

Attach the flaring tool to the end of the pipe and flare the metal.

valve in its body. Whether you are using propane or a blow torch, the procedure for sweat-soldering copper pipe or tubing is the same:

1. Brighten the end of the pipe with sandpaper or steel wool. You have to clean enough of the pipe for the fitting, plus about ½ inch more so that the solder will flow completely around the edge of the fitting. Also shine the inside of the fitting socket, using a clean pad of steel wool.

2. Spread a nonacid paste soldering flux on all of the brightened metal, that is, around the outside of the pipe and the inside of the fitting. When you heat the copper, the flux will prevent it from oxidizing before you get the solder on it.

3. Now put the fitting on the pipe and heat the metal with a propane torch. Keep your torch moving constantly around the fitting, and always hold the flame at right angles to the metal. You want to heat the fitting and about a half-inch of the tubing behind it.

4. When the tubing and fitting are hot enough to melt solder, hold the solder against the joint and slide it around the pipe. Actually, capillary action will suck the solder into and around the joint even if the joint is in a vertical position.

5. While the solder is still warm, wipe it with a soft cloth to remove excess solder and flux.

Flare Joints

One of the alternatives to sweat-soldering copper is to use flare joints, and there are two kinds of flaring tools you can use in making them. One is a knock-in type; the other operates on a screw principle, which is the more versatile of the two.

Flaring with a Screw-Type Flaring Tool

1. Slide the flare nut over the tube. It should face outward, toward the end of the tube.

2. Clamp the end of the tube in the flaring block. You should leave about ⅜ inch of the tube showing beyond the block.

3. Position the yoke of the flaring tool on the flaring block. Be careful to line up the cone with the center of the tubing. Lock the cone in position and turn its handle until the cone forces the end of the tube to flare outward.

4. Unscrew the flaring tool and remove it from the tubing.

5. Clean any burrs from the flare. Then place the fitting squarely against the flare and thread the flaring nut on the fitting. Tighten the nut with one open-end wrench holding the fitting and a second wrench on the flaring nut.

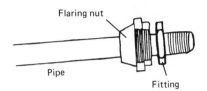

Slide the flare up to the end of the pipe and screw the fitting into it until it is hand-tight, then assemble the other half of the union and tighten the fitting with a wrench.

How to Flare

1. Insert the flaring nut over the end of the tubing. The nut must face outward.

2. Push the pointed end of the flaring tool into the end of the tubing and hammer it until the tube end is flared. You do not have to put the tube in a vise since you only need a few light taps with a hammer to produce a flare.

3. When the tubing is flared, assemble it to the fitting.

Using Compression Fittings

The third way of joining copper pipe or tubing is to use compression fittings. Compression and flare fittings look somewhat alike and both are more expensive than sweat-soldering, but at times their convenience makes up for the added cost. You may need to make a joint in limited space, for example, where it is difficult to get at the piping with a propane torch, or you may only need one or two joints in an already existing system. Once the tubing or pipe is cut and the burrs are removed, follow this procedure:

1. Put the compression nut around the tubing. It must face outward.

2. Slide the brass ferrule over the end of the tube.

3. Insert the tube into the compression fitting. Use one wrench on the nut and another on the fitting and tighten the nut until you squeeze the ferrule around the tube and form a watertight seal.

PLASTIC PIPE

There are several types of plastic pipe on the market today. Polyethylene (PE) is flexible, black, and is sold in long coils. It may be used for cold water service outside the house only. Polyvinyl chloride (PVC) is a rigid pipe suitable for both cold water lines and drainage systems. Chlorinated polyvinyl chloride (CPVC) is engineered to withstand 100 psi pressure at 180°F so that it can be used for both hot and cold water service. Acrylonitrile-butadiene-styrene (ABS) is black and used primarily in DWV systems. Polybutylene (PB) pipe is flexible thermoplastic. Colored beige or black, it offers you a choice between rigid CPVC and flexible PB when building a water supply system. Which one you use involves the same decisions as for flexible and rigid copper piping. Like CPVC, PB will withstand 100 psi pressure at 180°F; both CPVC and PB are accepted by the Federal Housing Administration for hot and cold water supply.

Besides being light in weight and easy to cut and assemble, plastic pipe also costs less than any other pipe. Its inner walls are extremely smooth so water flows easily through it. Consequently, any plastic pipe can handle the same flow of water as the next larger size metal pipe. Plastic also offers a natural insulation in that there is very little sweating in the cold water lines or heat loss in the hot water pipes.

Plastic does have some drawbacks. Only CPVC and PB pipe can handle the temperatures of hot water found in a home plumbing system. Any plastic pipe (as well as copper pipe) can be punctured by nails. Extremely high temperatures along with too much pressure can damage it, but that could happen only if a water heater's temperature and pressure valve failed. In that case, plastic water supply pipe would act as an alternate safety device, letting the heat and pressure escape before it could cause an explosion. Finally, if a joint in a plastic pipe system must be disassembled, you cannot just take it apart. You have to cut out the fitting and replace it with a new one, joining it with short lengths of pipe and couplings. The same must be done with a copper pipe from which the water cannot be drained to melt the soldered joints. With PB, however, which is joined mechanically, any joint allows take-apart and put-together at any time. In other words, every joint is a union.

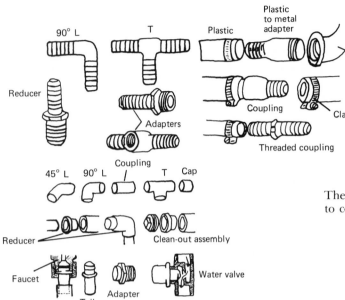

There is a wide range of fittings for plastic pipe, not only to connect plastic to plastic, but to metal as well.

Not all local plumbing codes permit the use of plastic piping, in spite of its many advantages. In the case of some large cities antiquated plumbing codes have yet to be updated to include plastic. Elsewhere, the use of plastic piping in water supply and DWV systems is readily accepted, although some municipalities have reservations about using it in hot water.

If you are planning to use plastic piping in your home, be certain the plumbing code in your locale permits your doing so, and only use whichever type is allowed by the code.

Rigid chlorinated polyvinyl chloride (CPVC) is designed specifically for use in water supply systems. You can purchase it in ¾-inch and ½-inch diameters that come in rigid 10-foot lengths. The fittings for CPVC are as varied as those for galvanized steel or copper and include adapters that allow you to connect CPVC safely to other pipe types.

Flexible polybutylene (PB) can also be used in both hot and cold water supply lines. It is made in ⅜-, ½-, and ¾-inch diameters, and is sold in 25-foot and 100-foot coils. PB is extremely flexible, so by using it you can avoid having to make a great many joints that might otherwise be necessary. It is ideal for remodeling work where the pipes must be snaked through walls.

Polyethylene (PE) tubing is made for connecting sinks and toilets to the water supply. It is sold in long flexible coils with diameters of ½, ¾, 1 inch, and larger. The fittings for PE include adapters with inside and outside threads, elbows, couplings, and Ts. When using these adapters, the PE pipe fits over a serrated end of the fitting and is held in place with a stainless steel pipe worm-drive clamp. A particularly good use for PE pipe is to bring up water from a well or water main. It is manufactured in several grades to withstand as much as 160 psi.

There are two plastic pipes designed for use in DWV systems, PVC and acrylonitrile-butadiene-styrene (ABS). The PVC used for DWV systems is beige, white, or gray, and the ABS is black. PVC is more resistant to solvents than ABS.

The plastic pipes and fittings used in DWV systems are all lightweight and require fewer supports than metal pipes. But the PVC offers one small advantage in that one of its two 3-inch nominal sizes—called Schedule 30, which is used for main soil stacks—has an outside fittings diameter of 3½ inches. This permits you to run the stack with fittings inside a 3½-inch-wide stud wall. The 3-inch Schedule 40 PVC and ABS DWV pipes have thicker walls than Schedule 30 PVC, so if you run them through a standard 2 × 4 stud wall you will have to furr out the wall an extra inch.

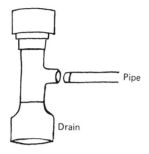

Pipe

Drain

The moment the solvent-weld is applied it begins to melt the plastic to such a degree that two pieces that fit tightly before they are painted will suddenly slide together very easily. Thirty seconds after they are together they are welded for all time.

While Schedule 30 is smaller than Schedule 40 on the outside, they are virtually the same size inside. Schedule 30 comes *only* in 3-inch PVC.

Joining Plastic Pipe

Whether you are joining PVC, ABS, or CPVC, the process is always the same; only the type of solvent used is different. These plastic pipes are solvent-welded at the joints. In other words, you glue a joint and once it is made, there is no way you can change it without cutting the fitting out of the pipe. Then you have to fill the space with a new fitting, plus short pipe lengths and couplings to make up the difference in lost pipe. Take the time to do it right the first time and you will have no trouble.

Solvent-Welding Plastic Pipe

1. When you measure plastic pipe, also measure the fittings to allow for make-up. The socket depths in plastic fittings vary, so measure carefully and add the depth to your face-to-face pipe length measurement.

2. You can cut any plastic pipe with a fine-toothed saw, but if you use a miter box, you can be sure of getting a straight, square edge. You can also cut it with a plastic tubing cutter. The burrs left on the inside of the pipe can be shaved off with a knife; you can sand smooth the rough outside edges.

3. Always test-fit the pipe in its fitting. A proper joint should be tight enough to keep the fitting from sliding off the pipe when you hold it upside down.

4. Clean both the pipe and its fitting with a cloth. If you are working with CPVC, wipe the solvent cement *primer* on the inside of the fitting and the outside of the pipe. Let dry and then apply the cement. (The solvents used with all other plastic pipe do not require a primer, but some pipe manufacturers recommend that one be used anyway). The solvent-welding cement is painted *liberally* on the pipe and inside its fitting. You have less than a minute before the cement dries, so work quickly.

5. As soon as the solvent is applied, push the fitting over the pipe and twist it slightly until it bottoms. Then adjust the fitting quickly so it is aimed in the right direction. Hold the joint together for ten seconds. If the joint is proper you will have a bead of dissolved plastic around the outer edge of the fitting.

The solvent will set in thirty seconds, but try not to move the joint for at least three minutes. It will be strong enough to withstand water pressure after an hour, but it is best to wait a minimum of sixteen hours before you pressure-test it. Be sure to use a quality solvent cement. Cheap cements are poor insurance against leaks.

Rigid plastic pipe can be joined to copper tubing by flaring and the use of a flare coupling. To join it to galvanized steel, you need a fitting called an adapter, or a transition adapter.

PE pipe is jointed to threaded pipe with a threaded-serrated adapter. It has one end threaded to fit the steel pipe, and the other end made for solvent-welding. Use of the transition adapter is recommended for pressurized hot-water connections. It has a thick rubber gasket between the two materials to take thermal movements without leaking. You can push the serrated end into the plastic pipe and hold it in place with a worm-drive pipe clamp.

CAST-IRON PIPE

Cast-iron is the most permanent DWV system available, and has long been a stand-by for the main stack and the waste disposal line. The pipe can be purchased in 2-, 3-, or 4-inch diameters and in 5-

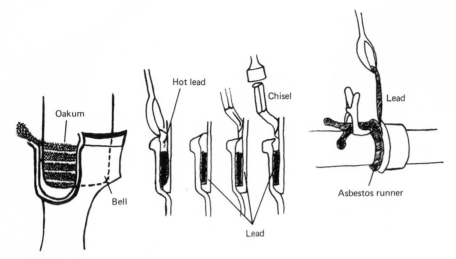

The spigot must be stuffed with oakum and then covered with molten lead.

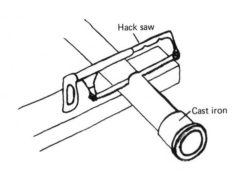

To cut cast-iron pipe, first saw a $1/16$-inch-deep groove around the pipe with a hacksaw . . .

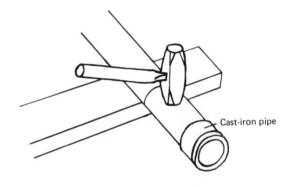

. . . then hammer the pipe until it breaks apart.

or 10-foot lengths. The cast-iron pipe used in waste lines has a bell-shaped hub at one end and a ridge known as a spigot at the other. The spigot fits loosely around the hub and the joint must be caulked with a loose fiber-filler called oakum. The oakum is tamped down around the hub and fills the space between it and the spigot to within an inch of the top. Molten lead is then poured over the oakum and fills up the last inch of space around it. The hub always faces the direction of water flow so that no waste will catch in the joint.

The work involved in joining cast iron is not difficult, but it requires a large heating unit, a ladle, and plenty of lead. So while you can probably rent the necessary tools from a plumbing supply store, assembling cast iron is usually left to the professional plumbers who already own the proper

equipment. If the pipe runs horizontally, by the way, it should be supported every four feet as well as at each joint.

Procedure for Cutting Cast-Iron Pipe

Formidable as it looks and feels, cast iron is relatively easy to cut; cutting cast iron takes more time and patience than skill. Saw a $1/16$-inch-deep groove completely around the pipe with a hacksaw. Now tap the pipe all the way around its circumference with a hammer and cold chisel. Keep tapping until the pipe gives a hollow, dead sound, then strike the pipe near the cut until the waste piece falls off. If you have a lot of pipe to cut, you can also rent a special chain-type pipe cutter that will make the job considerably quicker.

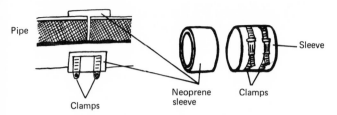

No-hub pipe is assembled by sliding a neoprene rubber sleeve over the ends of the pipe and securing it with clamps.

As an alternative to cast-iron pipe, most plumbing codes will allow either copper or plastic pipe, in which case you can construct a main stack out of either 3-inch or 4-inch-diameter pipe. The Cast Iron Soil Pipe Institute has also approved the use of No-Hub cast iron for use in drainage systems. No-Hub pipes have no bells or spigots, but are straight pieces of heavy cast-iron pipe.

Joining No-Hub Pipe

No-Hub pipe is butt-joined, so when you measure it, you do no need to allow any make-up distances for the fittings. Cut the pipe to the exact lengths you need and place them end to end. You can use all of your cut-off pieces and if ever a joint must be changed it is easily opened. The joint consists of a neoprene rubber sleeve, stainless steel shield, and a pair of clamps. Place the rubber sleeve over one pipe and the stainless steel shield around the other one. Now bring the pipes together, working the shielded pipe inside the rubber sleeve until it bottoms against the narrow separator ring around the center of the sleeve. Slide the steel shield over the sleeve and alternately snug up the screws on the two clamps.

5 Roughing-In

You must be prepared to take on some basic carpentry if you intend to do any extensive plumbing beyond repairing the fixtures in your home. If you plan to install some plumbing in an add-on room, or construct an extra bathroom, or any project that involves extending the house water supply and DWV systems, the work you do in running pipes through the walls, floors, and ceilings is known as roughing-in. In spite of its name, roughing-in demands some very careful work and also has some rather inflexible rules that must be followed for the sake of safety.

PLANNING

The cost of pipe and its fittings is high, so when you are planning a plumbing system think in terms of running your pipes as directly as possible to their destination. The DWV system requires larger, and therefore more expensive piping, so it should take precedence in your thinking and be constructed first. Moreover, if you plan carefully and cut a large enough way through the house for the DWV, you will be able to use the same route for running the smaller water supply pipes and save yourself considerable time and labor. But you must not weaken your house structure excessively in plumbing it.

The DWV system includes a 4-inch cast-iron house drain and main stack, although many local plumbing codes permit easier-to-install 3-inch to 3½-inch copper or ABS or PVC plastic pipe. The main drains and vents must be either 3-inch or 4-inch diameter pipe if they are serving toilets or a group of fixtures. Shower drains are 2 inches in diameter while 1½-inch pipe is used for sinks, bathtubs, laundry tubs, clothes washers, and dishwashers. While lavatories used to have 1¼-inch drainpipes, modern installations are virtually all 1½-inch pipe.

As a rule, the water supply system begins with a ¾-inch or 1-inch diameter service entrance; the hot and cold water mains that travel from room to room are ¾ inch in diameter. The branch lines that carry water from the mains to the fixtures are normally ½-inch pipe, with the exception of toilets, which use ⅜-inch pipe. Sometimes toilets are plumbed with ½-inch pipe to save bothering with another pipe size for this one fixture.

When you are making additions to existing plumbing, you are likely not to have as many options for pipe location as you would if you were installing a system in a house under construction. However, you can cut into existing lines by using a T or unions to make your connections. You can also connect the vents for new fixtures to the existing system by means of revents. It is best to consult your local plumbing code as to exactly how far a given fixture can be situated from the main house vent.

CARPENTER'S TOOLS

When roughing-in you will need to cut through a considerable number of joists, studs, and floors; so an assortment of chisels, hammers, and hand saws are useful to have, although in many instances either a saber saw or circular power saw will make your cutting easier. An electric drill, together with spade bits and a hole saw attachment is practially mandatory. Drilling is neater, quicker, and does not weaken studs and joists as much as notching them for your pipe runs.

OF NOTCHES AND HOLES

Pipes should be as unobtrusive as possible, which amounts to saying they must travel under floors, over ceilings, and up the insides of walls. Inevitably, if you intend to run a pipe through your house so that it is not readily visible, you will encounter joists and studs that you must either go over, under, or, preferably, through. But the moment you remove a chunk of wood from any framing member, you are weakening it considerably some specific rules about notching and drilling that must be observed. The penalty for ignoring these rules is to risk having your house fall down.

Rules for Cutting Holes

Whenever possible, drill through a joist or stud rather than notch it. In fact, your local plumbing code may not allow you to do anything but drill, in which case you will have to assemble your run in short pieces. Aside from any specifics set forth in the plumbing code, bear these rules in mind as well:

1. You can drill a hole anywhere along the length of a joist. But center the hole between the top and bottom of the joist and keep it at least 2 inches away from any edge.

2. The diameter of the hole should be less than one quarter of the width of the joist. For example, if you are drilling through a 2 × 8 joist, the largest

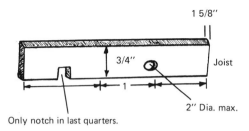

Only notch in last quarters.

There are some specific rules to follow about where you can safely drill or notch a wooden frame member. See text.

hole you can drill is 2 inches (one quarter of 8 inches).

Rules for Notching

There will be situations when notching is the only way of getting past a stud or joist. For example, a rigid pipe cannot be fed into a series of drilled holes in framing. If you absolutely must notch, at least follow these guidelines:

1. Only notch in the last quarter—the end quarter—of the joist. That means there should be no notches in the middle two quarters of the length of the joist.

2. Never cut deeper than one quarter the width of the joist. Thus, a 2 × 8 joist should be notched to a depth of no more than 2 inches, the less the better.

3. Nail a steel plate or strap, or a piece of 2 × 2 wood across the open end of the notch to brace the member. Nail on 2 × 4 or 2 × 6 lumber on either side of the member at the notch for even stronger support.

Rules for Notching Studs

It is a little safer to notch a stud than a joist, if only because the grain of the wood is running vertically, but there are rules for cutting studs as well:

1. Never notch a stud deeper than two-thirds its width. Thus, a 2 × 4 stud, which in reality measures about 3¾ inches in width, can support a notch 2½ inches deep.

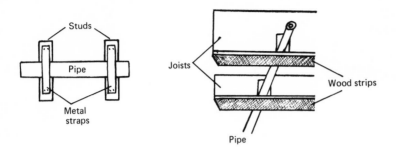

Notches should be supported by metal straps or wooden braces.

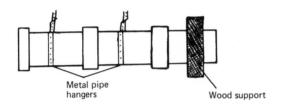

Pipes can also be hung from joists using perforated metal strap hangers.

2. If the notch is situated in the bottom half of the stud and is deeper than one-third the width of the wood, it must be braced with a steel strap or piece of wood nailed across its face. The steel, by the way, helps protect the pipe from errant nails.

3. A non-load-bearing wall is usually any wall that is not around the perimeter of the house or directly over the main support beam that runs the length of the middle of your cellar. You may notch the upper half of a non-load-bearing wall stud to half its width, providing that there are at least two unnotched studs beside it. Even so, it is recommended that you notch no more than two studs in a row.

Obviously, if you are notching a stud, you must first remove whatever wall covering is nailed to it. The procedure for notching is to make two cuts in the stud or joist that are of the proper depth and with enough space between them to accept the pipe. Then punch out the wood between the cuts with a hammer and chisel.

Some alternatives to drilling or notching joists are to cross your pipes under the joists and hang them with perforated metal straps cut to length and nailed to the side of the joist. Whenever possi-

ble, run pipes between the joists and support them on wooden braces nailed between the joists. You may also be able to lift a floorboard and cut through the subflooring to make a partial notch in the joists for your pipe, then replace the floorboard. The main stack is supported at its base by wooden members nailed into position under its last joint. The vertical run of the stack will be interrupted by both drainage and vent pipes that connect into it and help to support it.

PROBLEMS IN SPACE

Some partitions and many outside walls are too thin to allow you to run any pipes through them. When you are confronted by not enough space, your only solution is to build out the wall by nailing anything from 1 × 2 furring strips to the edges of the studs to framing the entire wall with 2 × 8s, running your pipes, then resurfacing the wall. The reason you have to go through all this carpentry work is that pipes and their fittings take up space. They need to be able to shift as the building settles, and so that vibrations or radical temperature

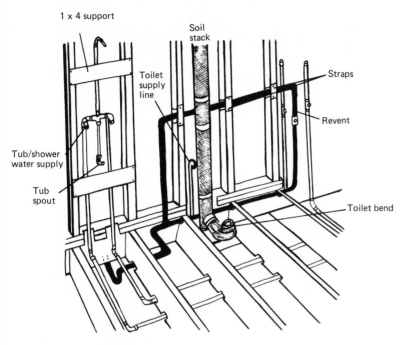

1 x 4 support

Soil stack

Toilet supply line

Straps

Revent

Tub/shower water supply

Tub spout

Toilet bend

This is what the inside of your bathroom walls and floor looks like.

changes do not cause either air or water leakage somewhere in the walls long after the pipes have been closed up and the rupture is difficult to locate.

Bear in mind that fittings are always a little larger than the pipe. A 2 × 4 stud is roughly 3½ inches wide, which is not enough to take a 2-inch cast-iron pipe with hubs. It can accept a 3-inch copper, plastic, or galvanized pipe and their fittings. You can fit a 2-inch cast-iron pipe between 2 × 6 joists, which have 5½ inches of clearance; but in order to hide a 4-inch cast-iron main stack, you will need to stand it between 2 × 8 studs.

You can run a pipe through any space that is roughly ½ inch larger than the nominal diameter of the pipe, but you will have to allow still more room at whatever point you are installing the fittings. That extra space must be increased even more if you are using threaded pipe, so that you have enough room to tighten your connections.

6 Installing a DWV System

Whether you are installing the entire plumbing in a new house, extending the existing pipes to a new room, or just making a few minor additions to your system, your work begins with the DWV. Actually, the work begins with your local plumbing code, which sets forth the tolerances and materials you can use as well as the dos and don'ts of whatever you are planning.

Assembly of the DWV normally begins by positioning the toilet drains and then working back to them from the sewer or the building drain. If you have only one toilet, it must drain directly into the main stack. Additional toilets may connect to the stack via branch drains, providing they are no farther away from the stack than is allowed by your local plumbing code, which is probably a distance of no more than 16 inches. All drains must slope at least ⅛ inch for every foot the drain runs. The branch drain must also be the same diameter as the main stack and so will be either 3 inches or 4 inches in diameter.

If your secondary toilets are too far removed from the main stack (more than 16 inches) they must be given a stack of their own that is the same diameter as the main stack, with at least a 2-inch pipe used in the vent portion of the stack. Normally, the toilet's vent is 3 inches or 4 inches. Every secondary stack must be vented through the roof.

POSITIONING THE TOILET DRAIN

The lead closet bend that connects to the toilet flange on the floor beneath a toilet presents some unusual problems. Try to position the toilet between joists so that they do not have to be notched; a large 3-inch or 4-inch notch in any joist will weaken it considerably. The distance from the finished wall to the center of the toilet flange is almost always 12 inches, but measure the toilet you intend to install, just to be sure.

Once you have located where the toilet will be, cut an 8-inch-wide strip of the flooring from the center line of the toilet bowl outlet to the wall. When you finish sawing through the floorboards you will have a hole 8 inches wide by 12 inches long that leads to the wall behind the toilet. But keep sawing until you have removed the 2 × 4 partition sole plate and given yourself enough space to install a T fitting for the stack as well as the closet bend assembly. If you are using 3-inch copper, 3-inch No-Hub cast iron, or Schedule 30 or Schedule 40 PVC or ABS plastic pipe, it will fit inside a standard 2 × 4 stud wall (the pipe but not the fittings). The only 3-inch fittings that can go within a standard wall are Schedule 30 PVC. Three-inch PVC or ABS plastic pipe with fittings

56

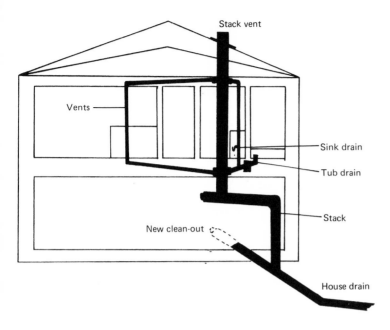

Stack vent

Vents

Sink drain

Tub drain

Stack

New clean-out

House drain

The DWV lines throughout any house must eventually connect to the house drain at the bottom of the building.

needs a 4½-inch-thick wall and 4-inch cast iron requires a wall made with 2×8 studs.

If you are working in a room that does not have a finished ceiling below it, such as on the ground floor, you need only cut a hole in the floor for the closet bend, and a second hole in the sole plate of the wall behind it for the stack. Your assembly work can then be done from underneath the floor.

If you must notch a joist to accommodate the bend or stack, make it only as deep as necessary and brace the joist on both sides with 2-inch-thick lumber. If a whole section of the joist must be removed, support each of the severed ends by nailing two 2-inch-thick boards that are the same width as the joists, across the faces of the joist to form double headers. The headers will be about 32 inches long and nailed to the joists on each side of the cut member, as well as to the ends of the severed joist.

When the holes are cut, preassemble the toilet drain. If you are working with ABS or PVC pipe you will need a T for the stack, two short lengths of pipe, and a closet bend, an elbow, actually. With cast iron you only need a T and the closet bend; any extra lengths of pipe you need can be broken off the bend. Copper pipe requires a T, two lengths of pipe, and a ¼-inch bend fitting, rather

than a closet bend. The plastic drainpipes are by far the easiest to work with since they are lighter and do not require a lead bend that must be sweat-soldered to make it watertight.

In fact, one manufacturer of PVC DWV pipe and fittings offers what is called a special waste and vent fitting that greatly simplifies a bathroom drainage hookup. Taking the place of the below-floor T, it provides side tappings at a 45° angle for both the lavatory and tub/shower drains. This saves using separate fittings not created to connect them. It also allows them to be wet-vented because the fittings are designed so that their drains connect above the toilet's drain. A cap at the top of the special waste and vent fitting lets you run the toilet vent stack on up through the wall. The fitting is made in 3- or 4-inch Schedule 40 and 3-inch Schedule 30. Side tappings are available in either 1½ inch or 2 inches.

For the purpose of preassembling the toilet drain, secure the toilet flange in its hole in the floor, and connect it to the bend using whatever length of pipe will position the bend at the proper height to connect to the stack. Hold the T for the stack in position and measure the distance from it to the bend. Cut the two pieces of pipe to their proper length and lay the entire assembly on the floor. Make certain the front of the bend and the back of the T are parallel. When you are sure of your measurements, assemble the pipe and fittings using either sweat-soldering (for copper) or solvent-welding (for the plastic pipes). Do not attach the flange, since this must be connected only when it can rest on the finished floor. You can hang the assembly in place, however, by bracing it with pieces of wood nailed to the joists.

When working with cast iron, follow the same process of preassembly. Measure and preassemble the T, ¼ bend, and pipe; then brace it in position using boards nailed to the joists. If you are using standard cast iron, all your connections will have to be caulked with oakum and wiped with molten lead.

INSTALLING A HOUSE DRAIN

With the toilet drain held firmly in place, drop a plumb line through the middle of the main stack T to within a fraction of an inch above the basement floor. If the T is on a second or third floor, you will have to lower the plumb until it nearly touches the first obstacle, then saw out a space large enough for the stack to go through the floor and lower the plumb to its next obstacle. When the plumb can reach from the toilet position to the basement, mark the floor beneath the plumb bob. This is the point where both the main stack and the building drain will begin. Using a chalk line, draw a line to the point in the foundation wall where the building drain will exit the house.

If the house drain is to be under the basement, dig a 2-foot-wide trench in the basement floor, using the chalk line as its center. The trench should begin 1 foot below the finished floor at the point where the stack begins, and slope downward ¼ inch for every foot of its run. When the trench is dug and graded, tamp the earth along its bottom to make it as firm as possible.

The building drain should have a cleanout assembly at the point where the main stack connects to it. The assembly consists of a cleanout T and two ⅛ bends or, if the drain is plastic, a Y with a cleanout adapter, plug, and a ⅛ bend. If the drain is cast iron, the cleanout assembly can include either a Y with a cleanout ferrule, plug, and a ⅛ bend; or it can be a T and two ⅛ bends. Whatever you are using, assemble the cleanout first and position it in the trench so that the plumb bob falls

in the center of the open half of the Y. Brace it with wooden members, then pour concrete around and under it so that it will remain in place. Wait until the concrete has hardened before you connect the building drainpipes to the cleanout assembly. If you are planning on having any branch drains, secondary stacks, or floor drains, connect to the building drain and place Y fittings at the appropriate positions along the pipe run to accommodate them. Actually, you might as well build all of these secondary drains now, so that the basement floor can be completed without having to wait until you finish installing the plumbing system. Whether the drain is cast iron or plastic, it should be leak-tested before covering.

When the drain line reaches its point of exit from the house, bore through the foundation wall and continue your sloped trench at least 5 feet beyond the foundation wall. It is at that point that the drain connects to your sewer line.

SUSPENDED DRAINS

Many homes have their house drain suspended above the basement floor. There is less hard labor to assembling a drain in this fashion, but which method you use depends on whether the drain connects to a municipal sewer system or enters a cesspool or septic tank in your backyard and how deep in the ground that connection must be made.

If you are suspending a drainpipe it can be 3-inch plastic or copper, but must be 4 inches in diameter if it is cast iron. The pipes are assembled with a Y or T fitting and cleanout plug at the base of the stack and a second cleanout assembly positioned just before the drain leaves the basement. If you elect to have only one cleanout plug, it should be at the base of the stack and the code requires that the pipe be 4 inches in diameter.

You may also need a reducer fitting to connect the cleanout assembly with the stack if, for example, your suspended drain is 4 inches and the stack is 3 inches in diameter. From the cleanout assembly, establish a chalk line to the point where the house drain will leave the building and assemble the pipes, permanently bracing the line every 5 feet (or every 3 feet for plastic). The bracing can be done by inserting pieces of 2×4 stock between the pipe and the floor, or by hanging the pipe from

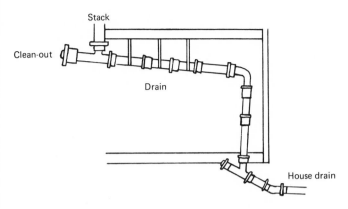

Whether it is suspended or buried, the house drain as well as the stack should have a cleanout assembly whenever it changes directions.

pipe hangers. The pipe hangers may be either metal straps or iron yokes attached to threaded rods that are fixed to the overhead framing members. Remember to include whatever Ys or Ts are necessary to let you connect additional fixtures to the drain if that is necessary. The end of the drain may have to turn vertically before it can leave the house. Avoid making a 90° turn; use 45° T–Y fittings instead.

STACKING THE STACK

When the house drain is in position, build your main stack from the cleanout assembly to the toilet assembly. If you are using No-Hub cast iron, copper, or plastic, you can join the pipe as you go. With standard cast iron, insert the spigots in their hubs and completely assemble the stack, then caulk and lead all the joints when you are finished. The main stack should be 3-inch or 4-inch plastic, cast iron, or copper and you should insert whatever Ts or Ys are necessary to allow drain and vent connections to reach the bathtubs, showers, toilets, sinks, and lavatories you are planning to connect along the way. None of them may vent below a toilet connection. Bring all of your connections out of their walls and cap them until you are ready to attach the fixture drains to them.

As you build the stack up through the house, you have two options whenever you reach an obstacle that must be skirted. You can use two ¼ bends assembled in opposite directions to form an S, or

you can use two 1 bends similarly to offset the drainpipe.

As you approach the point where each of the fixtures must intersect with the main stack, attach whatever Ts or Ys you need to connect their drains. Each of these must be placed so that the drains leading from the fixtures to the stack will be sloped ⅛ inch for every foot of their run toward the main stack. Their curving inlets should aim downward.

Building Up and Down from the Toilet Drain

As you erect the main stack up from the building drain to meet the toilet, bear in mind where the tub and lavatory drains will connect to the stack, and install the appropriate Ts and Ys. The tub will probably connect to the stack below the toilet. Lavatory drains are customarily positioned with their drains 16 inches above the floor and the drain must slope downward ⅛ inch for every foot of its run to the main stack where it is connected by a T or Y fitting. It is permissible to connect a drain to the main stack below the toilet *only* if the fixture is vented separately.

VENTS, REVENTS, AND WET VENTS

When the main stack has reached the fitting for the highest drain in the house, it ceases to be a waste stack and becomes a vent that must continue up through the roof. The vent portion of the stack

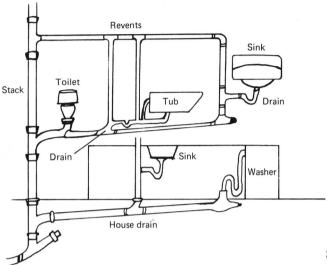

Some of the ways in which fixtures can be vented.

can usually be 2-inch pipe, although most often it is continued on up in the same size as its lower drainage portion. However, some local codes require that the vent be increased to 6 inches just below the roof, which means you will have to add a vent-increaser fitting a few inches below the roof, to prevent ice closure during cold weather. With cast iron, you must use special vent Ts, while plastic and copper pipe permit a sanitary T to be used, as long as it is inverted in the stack.

Every fixture drain must be vented, or else revented. If it is revented the vent line must rise from the trap at an angle between 45° and 90° and connect to the stack at least 6 inches above the flood rim of the fixture it serves to prevent any drain water from backing up the vent. If, however, there is a toilet on the second floor of your house, all of the first floor revents must tie into the main vent stack at a point above the highest fixture drain on the second floor. This usually means that most of your revents will run vertically up the inside of the walls and be connected somewhere above the second floor ceiling.

When you are installing a sink, lavatory, bathtub, or shower, try to locate it as close as you can to the main stack so that it can be drained and vented through the same pipe. That is, so you can use a wet vent. By definition of the National Plumbing Code, any part of a waste pipe that also serves as a vent for the fixture, is called a wet vent.

But to have a wet vent in accordance with the plumbing code you are limited to specific pipe lengths. The maximum distances for wet vents as permitted by the National Plumbing Code are:

1¼-inch drain pipe	3½ feet
1½-inch drain pipe	3½ feet
2-inch drain pipe	5 feet
3-inch drain pipe	6 feet
4-inch drain pipe	10 feet

(Longer distances—up to 8 feet—will work; however they will not meet the code.)

If the pipe diameters and their distances prove difficult, the fixture must have its drainpipe run down to the main stack under the floor, while the vent pipe rises vertically through the house until, somewhere near the roof, it crosses to connect with the vent stack (a revent), or alternatively goes through the roof and vents directly out of doors.

When you are extending a vent through the roof, bring the vent line up to the underside of the roof and then drill a hole through to the outside. The vent should be extended at least 6 inches above the roof, and no matter how tightly the vent line fits in the hole in your roof, it is not tight enough to keep water from dripping into the house. So the base of the pipe must be shielded with a flashing.

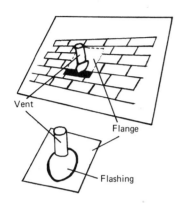

When installing any flashing over a vent pipe, use plenty of roofing cement or roof patch in all of the cracks and crevices.

INSTALLING FLASHING

You can purchase an angled, cone-shaped plate that comes with a rubber grommet around the narrow end of the cone to keep water from dripping down the pipe. To install the cone, pull the nails out of the shingles or other roofing material around the base of the pipe so that you can lift the roofing and place the base of the cone over the pipe and directly on the roof underlayment. Now liberally coat the base of the pipe with roofing cement or roof patch. Fill all the spaces you can see and then spread the tar on the roof under where the plate on the flashing will sit. Slide the cone down over the pipe and nail the plate to the roof underlayment using galvanized roofing nails. Spread cement over the base plate and around the grommet. If you have cemented the pipe heavily enough, cement will be oozing up under the grommet, but cover the outside of it anyway. Now nail the bottom course of roofing material over the base plate and coat the top of it with cement; at least coat the nail heads. Continue cementing and nailing the roofing material until all of it is back in place over the base of the flashing.

SECONDARY STACKS

Running revent pipes all over your attic can get to be both expensive and impractical, so it may be the better part of economy to build a secondary stack. The secondary stack need not service a toilet, but whether it attaches to a branch drain in the cellar or connects directly to the building drain, position a cleanout assembly at whatever point it changes directions into a horizontal run.

When planning a secondary stack, you can be less precise about its exact position if there is no toilet attached to it, but the rough-in work should be done beforehand and a plumb line hung down the way you have cut through to the basement. If there is to be a toilet, assemble and position the closet bend and hang a plumb line. Then assemble the cleanout assembly and center it under the plumb bob. Build the branch drain to the house drain (if that is necessary) and then build the secondary stack up to, and out through, the roof. As you go along, assemble connections for each of the necessary fixture drains and vents, and lead them out of the walls and cap them. You will need flashing at the roof, the same as for the main stack.

TESTING THE DWV

Before you close up any walls or ceilings, you should make sure all the connections in your DWV system are watertight. To test the DWV system you will need to rent a set of rubber plugs from a plumbing supplier to close off the closet bend and the sewer line at the cleanout assembly positioned near where the house drain leaves the building. Insert the plugs in the closet bend(s) and house drain cleanout assembly; then go up on the roof with your garden hose and pour water down the main stack until the stack is full. Wait about half an hour and then closely inspect all your joints for leakage. If you find any leaks, tighten the joint. This is the time to call the building inspector, if one is involved. If you don't see any water anywhere, pull the plugs and go about closing up the walls. If you have used ABS or PVC plastic pipe, beware that while the solvent-welded joints will be impossible to move three minutes after assembly, they should not be pressure-tested for at least 16 hours.

BASEMENT DRAINS

If your basement collects considerable amounts of water every time it rains, and if your house drain empties into a municipal sewer system, you can

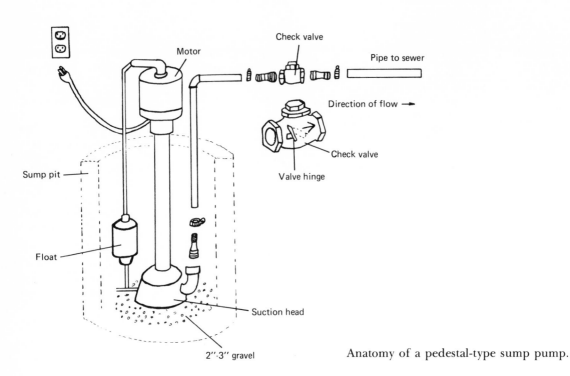

Anatomy of a pedestal-type sump pump.

construct a basement floor drain and connect it directly to the house drain. However, if the house drain leads to a septic tank, you do not want to flood it with excess water. So, the basement floor drain must be led to a separate disposal area, such as a stream or a dry well. In any case, make certain that the drainage area is large enough to handle the water demanded of it.

Connecting a basement floor drain to your house drain can present several different problems. If the building drain is below the basement floor, dig out the area for your drain and run a branch line directly to the house drain. Each floor drain then must contain a trap. But if the house drain is suspended, the basement floor drain should be graded down to a sump pit. The sump pit is actually a hole in the basement floor made of 24-inch diameter sewer tile stood vertically on a gravel base. Install a sump pump in the pit; it will automatically pump any water that collects from the basement floor up to the house drain.

SUMP PUMPS

Sump, or basement drainage, pumps come in a variety of designs, but all of them divide into either submersible or pedestal types. Whichever type you use, the pump will start automatically as soon as water in the pit reaches a predesignated level, and it will continue pumping until the pit is almost dry. The pump is typically controlled by a float that rises and falls with the water level in the sump pit and expels water through a pipe connected to the house sewer or some other convenient place.

The major difference between the submersible and pedestal pumps is the location of their motors. The submersible type has a completely water-sealed motor that resides inside the sump pit. With the pedestal type, the motor stands on a long shaft so that the motor is always completely out of the water.

When you are installing a sump pump, connect a 1¼-inch flexible plastic pipe to the pump's outlet (discharge) port with a plastic-to-brass adapter. The flexible pipe can lead to a sewer, the house drain, a slop sink, storm drain, seepage pit, or outside to the ground; but install a 1¼-inch brass check valve in the pipe somewhere near the pump. The valve should be connected with the arrow embossed on its side pointing *away* from the sump so that water cannot flow back into the pump. The pump should be plugged into a 117-volt, three-wire grounded outlet. Check your code.

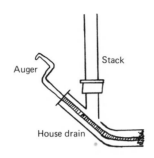

There should be a cleanout plug at the base of the stack and at each point the main house drain changes directions.

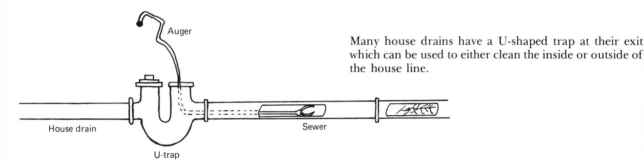

Many house drains have a U-shaped trap at their exit which can be used to either clean the inside or outside of the house line.

Some will not permit pump discharge into a house drain.

Sump pumps are constructed to give years of troublefree service, but since they usually operate in damp conditions, inspect the unit occasionally to make sure corrosion is not building up on any of its parts. Also check the motor and impeller bearings at least once a year to be certain they have adequate lubrication. When adding new lubrication, first drain and wash out the bearings. Keep the sump pit clean.

DRAIN AND WASTE SERVICING REPAIRS

Since the DWV pipes are rarely under any great pressure, they are not likely to burst. After many years, metal ones may corrode or rust and have to be replaced, but most likely the only repairs you will ever have to do on them is the job of unclogging. Usually, if waste backs up in any of the sinks or lavatories in your house, the clog is between the

sink drain and the stack, and you can reach it by running a snake down the fixture drain or through its trap.

Once in a while, the stack or house drain will become clogged, which is why all those cleanout plugs appear at the higher end of horizontal runs. The cleanout plug at the base of the stack is angled toward the house drain; the chances are that if the stack is clogged and you have not been able to reach the obstruction from any of the fixtures, the blockage is in the house drain. Using a pipe wrench, unscrew the cleanout plug and insert either a tape or coil spring auger into the drainpipe. The most practical size auger for cleaning house drains is the flat ribbon type with a ½-inch-wide bulbous head. Keep pushing the snake into the drain until it stops; then withdraw it a few inches and ram it forward. Keep pounding the obstruction until you break through it. Then flush all of the toilets in the house several times to wash out the remains of the blockage.

If you cannot reach the blockage from the stack end of the house drain, try getting at it from

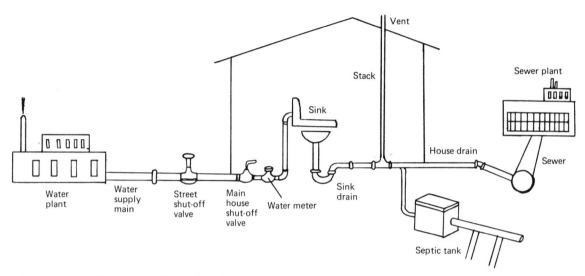

The water supply–sewage disposal cycle.

the house trap side, if there is a trap. House traps are often U-shaped fittings with covers in the top of both legs of the U. By opening the cap nearest the main stack, you can run your snake back toward the stack. If the house drain is free of obstacles, open the street-side cap and push your auger toward the outside until you reach the blockage.

If you find you cannot free a blockage in your underground sewer line, it may be that tree roots have forced their way into tiny cracks in the sewer line. If your house is connected to a municipal sewage disposal system, you might possibly get the city to clear the blockage. But if the offending tree is on your property, it is likely to be your responsiblity to clear the line.

The way to clear a sewer line blocked by tree roots is simple: Call one of the experts listed in the yellow pages of your telephone directory. The professionals have an electrically powered auger with razor-sharp knives at the end, which they can drive into the sewer line from the house cleanout assembly. The auger can slice through tree roots or any other debris and clear the line. Having paid to learn that lesson, you can prevent the tree roots from blocking your sewer again by pouring a double handful of copper sulphate crystals into the main trap once a month. The crystals will effectively inhibit root growth in the sewer. Copper sulphate is poison, so be careful with it; store it where children cannot reach it.

DRAINAGE OUTSIDE YOUR HOUSE

The waste water that leaves your house drain is disposed of either collectively, or privately. The collective system is usually operated by a municipality and consists of sewer pipes leading from numerous houses into trunk lines that eventually terminate at a sewage-treatment plant. Modern sewage-treatment plants remove any solids in the sewage by pumping it through a succession of screens. The solids are collected in storage tanks where it is dried and can then be used as fertilizer. It can also be used to create a gas that will operate the huge generators necessary to run the water treatment plant. In the meantime, the sewage water is going through a series of settling, aerating, and chlorinating processes that purifies it until, in some cases, it is clean enough to drink. At which point it is discharged into the nearest waterway or into the ocean.

Private sewage-disposal systems are nowhere near as hygienic nor so efficient as a collective system, but thousands of homes in the United States still rely on them to get rid of their waste water and solids. Most private disposal systems begin with a septic tank, which slowly breaks down the solids of anaerobic bacterial action. The solids settle in the bottom of the tank, which must be cleaned out every several years or so. The liquid effluent coming from the tank drains into a net-

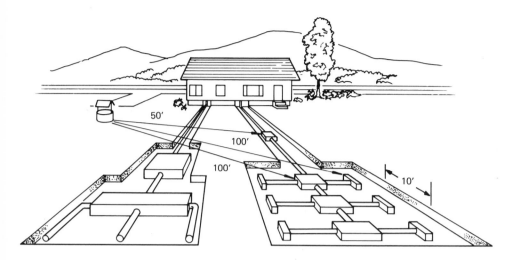

There are explicit minimum distances that must be maintained between a disposal system and fresh water supply sources.

work of underground trenches or pits where it seeps into the soil. Sooner or later, the disposal field may become clogged by the minute solids suspended in the effluent and must then be enlarged.

LAYING OUT A SEPTIC TANK AND DISPOSAL FIELD

Anytime you are contemplating laying out a disposal field, begin by reading your local plumbing code. You will find that the codes govern distances from the disposal area to your house, wells, property lines, and trees; and they should be adhered to strictly.

The next order of business is to determine how much effluent the ground can absorb, by conducting a percolation test. You perform the test by randomly digging six holes in the area where you intend to put the disposal field. The holes should be about one foot in diameter and dug to whatever depth the absorption trenches will be, which is normally between 18 and 36 inches. Fill the holes with water and wait 24 hours. Now adjust the water level in each hole to a depth of 6 inches and stand a ruler in the water. Clock the number of minutes it takes the water level to go down one inch. The table shown here gives the minimum number of square feet in the absorption field

needed for each bedroom in your house. If, for example, you have three bedrooms and it takes 15 minutes for the water in the holes to drain off one inch, multiply 190 square feet times 3. The disposal field will have to be at least 570 square feet in area.

PERCOLATION TEST

Minutes for Water to Fall 1 inch	Square Feet of Disposal Area × Number of Bedrooms
2 (or less)	85
3	100
4	115
5	125
10	165
15	190
30	250
45	300
60	330
Over 60	Area not suitable for disposal

The septic tank is connected directly to the house drain via a sloping, watertight sewer pipe. The tank itself can be constructed from all kinds of

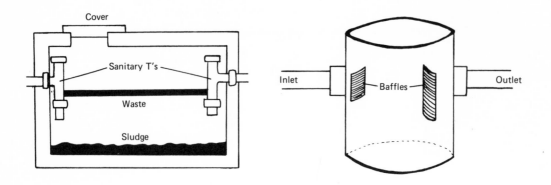

A precast concrete (left) and steel septic tank (right).

piping materials including precast concrete, brick, clay tile, concrete block, asphalt-coated steel, or redwood. The tank is completely buried in the ground and must be large enough to accommodate the household it serves. Local codes vary on the tank's capacity, depending on whether they allow you to use the tank only for toilet wastes, while leading all other wastes directly to the absorption field, or whether all wastes must go through the septic tank. The accompanying table assumes that the tank will handle all sewage coming from the house.

SEPTIC TANK CAPACITY

Number of Bedrooms	Minimum Tank Size
2	750 gallons
3	900
4	1000
5	1250

PUTTING IN A SEPTIC TANK AND ABSORPTION DISPOSAL FIELD

There is a lot of earth-moving that goes on before you can install a septic tank and disposal field. If you intend to dig all those trenches with a spade you are in for some long, hard hours of manual labor; hiring a contractor with the right heavy equipment, such as a backhoe, makes the trenching considerably easier. But whether you do the digging yourself or hire someone else to do it, there are still some rules to follow.

The tank and field should be arranged with their distribution boxes and seepage lines on a slope below the house and the location of every stream, well, property line, and pressurized pipe must be taken into consideration. Your local health or building department can not only advise you as to the design of the system, but may even send someone to help you determine the final layout of the field. In general, the more area you have to use for your disposal field, the better. Consider that you will need, at the very least, a quarter of an acre, or about 10,000 square feet.

If you hire a contractor to dig the septic tank, sewer line, and seepage trenches, mark out where you want him to dig with stakes driven in the ground. Unless he already knows, you must also tell him the depths and slopes you need to meet your health department code. The sewer line should slope at ⅛ inch per foot; if it angles steeper than that, at least be sure the last 10 feet leading up to the septic tank is no steeper than that. The seepage trenches can vary from dead level to a slope of 6 inches per 100 feet (but check your local requirements). Between 2 inches and 4 inches per 100 feet is a reasonable slope. Trench widths will vary from 12 inches to 36 inches, but the most common widths are 18 inches. The hole for the septic tank must be deep enough so that the top of the tank is at least one foot below the surface of the ground, although in colder climates you may be

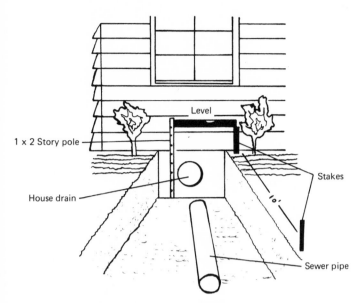

Establish the proper drainage slope by driving stakes 10 feet apart along the side of the sewer trench and using a story pole made from 1×2 stock.

required to cover the tank with two or three feet of eath to keep it from freezing.

Digging the Sewer

The building sewer is connected directly to your house drain, which technically ends somewhere around 5 feet outside the foundation wall. The sewer thus begins at the house drain and runs as far as the septic tank.

The pipe used for sewers depends on local building codes and could be cast iron, plastic, pitch-fiber, or vitrified clay with a minimum diameter of 4 inches. (Some codes permit a 3-inch sewer line.) If the code permits you to use plastic or pitch-fiber pipe, use one of them because they are lighter and easier to work with.

The sewer line must be watertight, particularly if it runs within 50 feet of a well or 10 feet from any drinking water supply line that is under pressure. So whatever pipe you use must have sealed joints. There should also be a cleanout installed at any point the line changes direction by 45° or more, or if the slope changes 22.5° or more. If you have to make that much of a bend in the line, do it with a pair of 45° ells or ¼ bends, rather than a T or elbow. Also, if the septic tank is

more than 20 feet from your building drain cleanout, it is a good idea to include a cleanout within 5 feet of the tank. Presuming the sewer is no more than 4 feet deep, you can make a cleanout simply by inserting a T in the line and running a vertical pipe up to ground level and capping it with a cleanout. If the sewer is deeper than 4 feet, you will have to construct a manhole around the cleanout.

Sewer lines should be graded at least 1 inch, and preferably ¼ inch per foot. You are allowed to make it steeper but, surprisingly, if a sewer is too steep solids may become stranded in the pipe and can cause obstructions.

Digging the Trench

Sewer pipes are supposed to be underground and in order to get them there you have to dig the straightest possible trench from the end of your building drain to the septic tank. The trench should be properly graded and deep enough so that the pipe will not freeze or be crushed if, for example, it runs under a driveway. The trick, of course, is how do you make it "properly graded"? The procedure for establishing a proper grade is this:

1. Drive a series of stakes into the ground 8 inches away from the width line of the trench, and exactly 10 feet apart. The first stake you drive should be against the side of the house and this becomes your point of reference. The last stake should be next to the house side of the septic tank.

2. Lay a straight, 10-foot-long board on top of the first and second stakes and adjust the number-two stake until the board is level. You can either saw it off, drive it deeper into the ground, or pull it a little farther out of the earth. Continue down the line of stakes, adjusting each one until they are all the proper height to form a level line from the house to the tank.

3. Now stand a straight piece of 2×2 stock vertically beside the house drain and place one end of your level on top of the first stake, with the other end against the 1×2. Adjust the level until it is true and mark the 1×2. Assuming you are grading the sewer ¼ inch for every foot, the second stake, which is 10 feet away, will represent a depth of $^{10}\!/_4$ inches, or 2½ inches deeper than the first stake. Measure 2½ inches above the line on the 1×2 stock and mark it. Continue measuring and marking the stock every 2½ inches for each stake you have driven. You have now turned the 1×2 into a very useful gadget, which carpenters call a story pole.

4. As you dig your trench past each stake, stop to stand the story pole on the bottom and hold your level between the top of the stake and the appropriate line on the pole. Tamp the earth in the bottom of the trench until it is firm and even enough to support the sewer pipe.

DOWN IN THE TRENCHES

Some local codes require the use of either standard or No-Hub cast-iron pipe that measures more than 4 inches in diameter. The standard cast-iron spigot and hub pipe requires a somewhat wider trench so that you will have room to work on the joints. Every standard joint must be sealed by tamping oakum around the spigot until the space is packed to within 1 inch of the hub rim. Then pour between ¾ inch and 1 inch of hot lead on top of the oakum. The lead must be firmly packed against all of the surfaces it touches. (See Chapter 4.) This is lots of work. It requires the use of an asbestos rope tool called a joint runner, and also requires a caulking tool.

There are distinct advantages if you are permitted to use plastic pipe, if only because it is lighter and easier to assemble. If you did not plumb your house with plastic too, begin your run of plastic pipe by attaching a metal-to-plastic adapter to the end of the house drain. Then solvent-weld whatever lengths of pipe are necessary to reach the septic tank, joining them with couplings. Dig out for the couplings so that the pipe is supported uniformly along its length, not suspended between its couplings. At the tank, you will need a cast-iron pipe adapter, which is leaded and caulked into the tank inlet.

Pitch-fiber pipe is as light and easily worked as plastic pipe. Again, begin at the house drain with an adapter for what is there, and build the sewer line with 10-foot lengths of pipe assembled with couplings. The ends of the pipe are tapered, and joints are made simply by hammering a tapered coupling over the pipe end. Place a block of wood against the coupling (never the end of the pipe) and pound it onto the pipe with a sledge hammer until the tapered end is seated against the recess in the coupling. The heat generated by pounding the couplings down over the pipe causes the pitch-fiber to heat just enough to weld the joints. So make sure the fittings are aimed in the right direction before you join them. They cannot be removed without sawing the pipe.

If you cut a pipe, the end can be tapered with a special tool clamped into the end of the pipe and rotated until its cutters have shaped the outside of the pipe into a cone. You can also connect two untapered pipe ends with a sleeve and special cement. The final joint between the fiber pipe and the septic tank inlet must be done by cementing an untapered pipe end into the inlet hub.

One end of the ⅛ and ¼ bends for fiber pipe, as well as the Ys, is tapered on its outside so that it can receive a coupling. The other ends of the fittings are shaped like a coupling to accept the end of a pipe. The pitch-fiber T fitting has all three of its open ends recessed to receive pipe.

When laying any pipe, remember it is critical that the slope remain constant, and the joints are larger than the pipe itself. Thus, you must scoop out enough earth under each joint so that the entire length of the pipe can lie flat in the bottom of the trench. The soil underneath the sewer should be well tamped. Backfill around the pipes should be rock-free. The trench is filled up to ground level with thoroughly packed soil once the sewer is completely assembled and the select backfill has been placed around it.

CESSPOOLS AND SEPTIC TANKS

The difference between a cesspool and a septic tank is that the cesspool is merely a collection tank that allows raw sewage to leach into the ground. The septic tank is designed to break down the sewage by bacterial action and is therefore considered more sanitary. An increasing number of communities have outlawed cesspools on the grounds that they are a health hazard. If a cesspool is located anywhere near a source of pure water, it will eventually contaminate the water and make it dangerous to drink. Moreover, sewage leaching out of the cesspool will ultimately reach the surface of the land, where it decomposes and causes a foul odor.

On the other hand, the septic tank takes in raw sewage from the house and holds it while some of the solids settle to its bottom and bacteria dissolves a little more of the waste, leaving a relatively clear liquid at the top of the tank. The effluent then flows out of the tank into the perforated pipes that make up the disposal field and is dispersed in the soil.

A brand-new septic tank does not need any particular treatment to start the bacterial action, and you can build one yourself using redwood, bricks, or clay tile. You can also buy a precast concrete or metal tank, which in the long run is probably a wiser course to take. Some tanks on the market are designed with two compartments and include air intakes, inlets, outlets, and baffles to prevent splashing, as well as access covers for cleaning. The most modern septic tanks have two compartments and may go for as long as a dozen years before the sludge needs to be cleaned out of them. One small advantage to buying a septic tank is that the dealer will not only deliver it to your sewage site, he will deposit it in the hole for you. Make sure you get the hole specifications from him before you start digging.

Septic tanks have an inlet that accepts the sewer line from the house, and an outlet that is slightly lower than the inlet to keep sewage from backing up into the sewer line. The outlet leads directly to the distribution field. The tank is completely sunk in the ground and covered with about a foot of earth, which you will have to remove

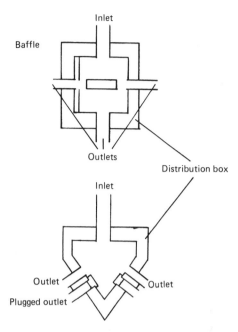

Distribution boxes come in all shapes and sizes and offer from one to several drainage outlets.

every year or so to open the access cover and check the sludge level in the tank.

Maintaining a Septic Tank

Depending on the number of people in the household, your septic tank should be uncovered and checked every year or year and a half. Open the access cover and push a stick to the bottom of the tank; if the sludge is 18 or 20 inches deep, the tank should be pumped clean. Don't be a hero: Call a professional septic cleaner to clean the tank and cart the sludge away.

Distribution Boxes

A watertight pipe runs from the septic tank outlet to a buried concrete vault known as the distribution box. You can build that, too, but it would be better if you purchased it from your local concrete products supplier. There are several distribution box designs, including square, round, and triangular; but they all have an inlet port and two or more outlets to divide effluent equally among the seepage runs. It is nothing more than a box in the ground; but in order to work effectively, you must be sure that it is placed so that all of its outlet pipes

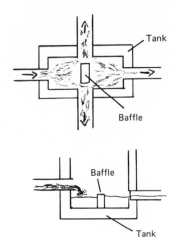

The distribution box must be placed with all its outlet pipes level with each other so that sewage will be distributed evenly between them.

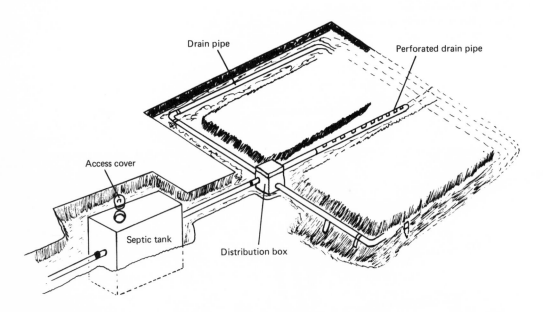

One of the many forms the pipes in a disposal field can take. A typical alternative is to lay the pipes in a continuous series of Ss.

are level with each other. You do that, of course, by leveling the top of the box in both directions.

The Disposal Field

The real trench digging begins about 5 feet beyond the distribution box. The first 5 feet of trench from the distribution box outlets will contain a watertight pipe that connects to the seepage pipes. Cut a series of parallel trenches 10 feet apart and connected by a trench at alternate ends, so that you create a series of right angled Ss. The trenches should be graded so that the seepage pipe will slope 2 inches for every 50 feet of their run. (Check your Code.) Dig as many trenches as you can in the area you have available to you for your disposal field. It is not absolutely necessary, but if the seepage lines are long enough, and far enough away from the septic tank, a distribution box can be placed in the center of the seepage pipes to help control the flow of effluent.

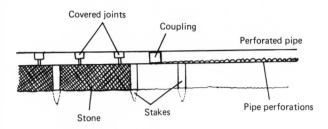

A pipe layout using tiles or perforated pipe.

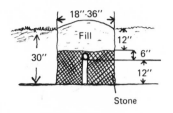

Cross section of a seepage pipe trench.

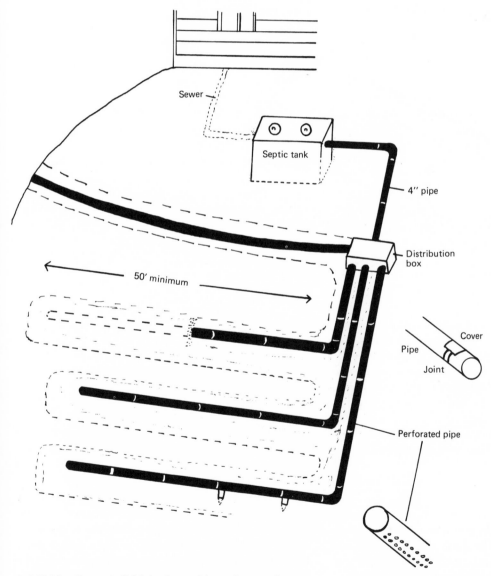

A hillside disposal field is dug with each trench at a higher elevation than the trench beyond it.

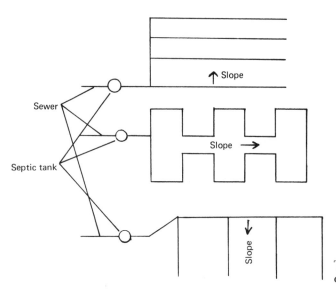

Three possible pipe configurations used in a sloping drain field.

The pipes you use for seepage should be 4 inches in diameter and can be made of concrete, clay, plastic (PVC or ABS), or fiber. Concrete and clay come in 1-foot lengths and are laid with ¼-inch spaces between them to allow seepage. The plastic and fiber pipes come in 10-foot lengths and have double rows of holes that are aimed *downward* when the pipes are laid. The perforated pipes are assembled with couplings that need not be made watertight except at turns. The linking trenches have watertight pipe at the end of each seepage pipe run. The free ends of any runs that do not have turns can be blocked with gravel or a large stone.

The trenches should be dug so that you have at least 12 inches of space above and below the pipe; they are normally not more than 3 feet deep. The bottom of the trench is filled with 12 inches of gravel. When you order the gravel, figure that one cubic yard equals about 1½ tons, and allow 10 percent for waste. When the trenches are bottomed with gravel, lay and grade the seepage pipes, then cover them with between ½ inch and 2½ inches of gravel. If you are using concrete or clay pipe, place a strip of 15-pound asphalt-saturated felt (tar paper) over each joint to prevent any of the backfill from clogging it. At this point, ask your local health or building department to inspect the work—it will avoid the possibility of having to dig up the pipes at a later date, just to get a Good Housekeeping Seal of Approval from your municipality.

Once you have clearance from the health department inspector, fill in the seepage trenches. First fill to 4 inches over the pipes with gravel. Then place a layer of felt over the stones, which will help to prevent soil from filtering down into the pipe joints. Finally, fill the last 12 inches of the trenches with soil. Do not tamp the earth, but mound it slightly above the surrounding terrain and allow it to settle. You can cover the trenches with any kind of landscaping except willow trees, which tend to send their roots into the seepage lines and clog it.

Steep Hills and Sloping Lands

If the land is exceptionally steep, you may have to construct a *serial* distribution field beyond the septic tank by cutting level trenches along the contours of the slope, with each trench lower down the hill than the one preceding it. The trenches are kept absolutely level so that the first trench and its gravel must be full to capacity before it overflows into the second, and the second must fill before it flows into the third, and so on.

If the land is hilly, or the available disposal area is too small for a trench arrangement, your final option is a seepage pit. With a seepage pit arrangement, the house drain leads into a septic tank, which is connected to a distribution box with watertight pipes. The distribution box leads to one or more seepage pits via loose-jointed pipes.

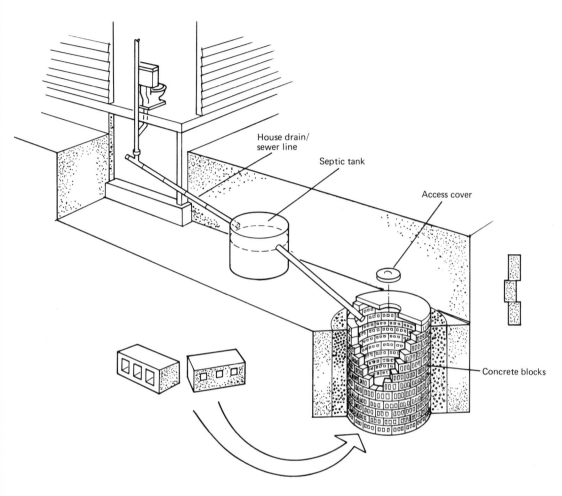

The last resort when your drainage field is very steep or rather small is the seepage pit.

The seepage pit, or pits, can be big enough to hold a bulldozer when they need to be cleaned out. They can also be considerably smaller. They can be round or any other shape, and they may or may not be walled. Sometimes a seepage pit is constructed with no walls so that nearby trees can fill it with their roots, which help to absorb the effluent in the pit. Normally, the pit is a hole dug in the ground and lined with loose-jointed concrete blocks, stones, or bricks and its dirt bottom covered with gravel. If there are no walls, or the walls are not inherently strong, the pit can be filled with stones. You can do just about anything with a seepage pit so long as it has some way of receiving effluent, and its sides and bottom are porous enough to then distribute sewage into the ground. In fact, if you are in love with the idea of seepage pits, you can put one or two of them at the end of any distribution field. The top of a hollow pit should have a removable access cover made of stone or concrete so that the pit can be inspected and cleaned from time to time.

Chemicals and Sewage Disposal

Aside from periodic inspection and cleaning of the septic tank and seepage pits, there is no maintenance to a private sewage disposal system. You can add chemicals to your sewage that will shorten the time needed for the solids to decompose; but

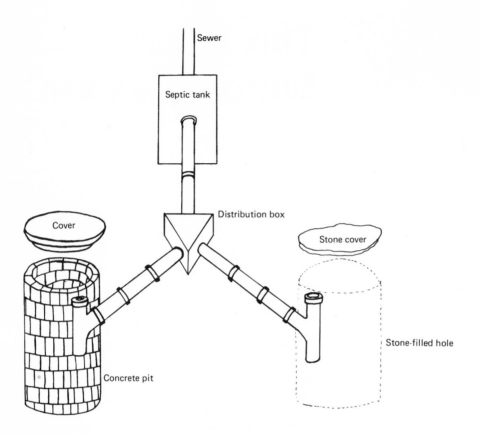

Seepage pits can be made of practically anything and can be arranged in almost any configuration.

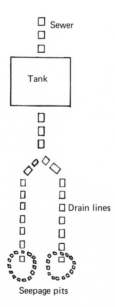

Seepage pits can, in fact, be added to the ends of a standard disposal field.

which chemicals you select depends on the sewage and local conditions and whatever regulations may have been established by your local health or building departments. Lime and sulphate of alumina are permissible in many communities as additives to sewage; and yeast can sometimes speed bacteriological action. About all you need is half a pound of brewer's yeast mixed in a pail of warm water, which is emptied down a toilet once every six months.

The only other maintenance involving a private sewage system is preventive: Try not to allow such things as cigarette filters, which will not decompose, into the system; they will only clog the lines. And avoid putting any unnecessary water, such as roof run-off or basement water, through the sewer. It will only serve to congest the absorption field that much sooner. If the earth in the field finally gets so filled that you can no longer drain your sewage into it, you will either have to dig a new distribution area or expand the one you have. With luck, that will not be for many years.

7 The Water Supply System

The water that enters your house begins as rain and reaches you via either a well or, for most homes, a municipal water supply system. It comes into the building under a pressure of between 60 and 100 pounds per square inch and immediately passes a main shutoff valve and then a water meter (if you are being charged for your water consumption). The entrance pipe is usually ¾ inch or 1 inch in diameter. Once inside the house it becomes the cold water main, which is normally ¾ inch in diameter. For the short run across your basement ceiling, the cold water main travels alone. Then it reaches the house water heating device, which it feeds into with a ¾-inch branch line.

The hot water main is a ¾-inch pipe that begins at the house heater and continues throughout the house to each of the fixtures in the building. In nearly every instance you will find the hot and cold water lines running parallel to each other. When the pipes run vertically for a distance of at least one floor, they are known as *risers*. The smaller pipes that lead away from the main lines to supply each fixture are known as *branch lines*.

Throughout the water supply system you will discover a myriad of valves, which show themselves as fittings with wheels or handles attached to them. Most are globe valves, and they operate like any globe-type faucet. They have packing nuts and packing, or packing washers, and stem washers at the bottom of their threaded stems. The reason you will spend very little time having to repair a leak in the valves is that for most of their lifetime they are left open, so their working parts do not have much opportunity to become worn. Then again, they may be left unused for so long that they become corroded and locked in their open position, which is annoying when that one time you want to close them comes around and you have to put a wrench on the handle before you can turn it. It should be noted that some valves, particularly the gate valve, which is often used as the main shutoff valve at the house service entrance, can take a long, long time to close. Gate valves are designed so that when they are open, the wedge at the bottom of the stem is completely retracted from the flow of water through the valve body. This makes them less resistant to water flow. Consequently, when you begin turning its handle you have a long thread to close before the gate is completely shut. Don't give up. Keep turning until the handle will not rotate anymore. At that point you can be sure that no water is flowing through the pipe.

Another problem that arises with the valves because of their long periods of disuse is that when you do turn one of them off and then on again, it may develop a leak around its packing nut. Tighten the nut about half a turn with a wrench and the leak should stop.

Valves are found throughout your water supply system. They can be seen guarding the cold water supply to your hot water heater, water treat-

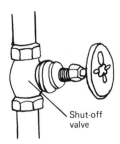

It is a good idea to tag each of the valves in your system so that every family member will be able to find the one they need during an emergency.

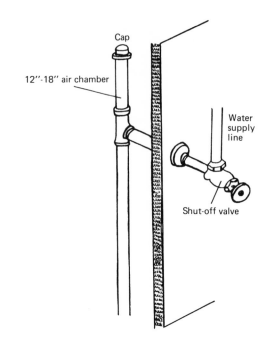

Every faucet should be supported by an air chamber.

ment equipment, at whatever point the hot and cold water mains turn vertically to rise up through the house, and every time a pipe emerges from the walls beside a fixture. Their purpose is always the same: to shut off the water supply to a particular area of the house in the event there is need to shut it off, while still allowing water to continue servicing the rest of the system without disruption.

Unlike the DWV pipes, water supply lines do not need to be of a large diameter, or even very heavy. They can be made of galvanized steel, brass, or copper. They can also be plastic, although some local codes—fewer every year—prohibit the use of plastic pipe for the hot water line.

Normally, you will find the water supply pipes beginning their runs through the house if you look up at the basement ceiling, where they are secured to the bottom of the joists with metal hangers. The mains are usually ¾ inch in diameter and the lines that branch off them are typically ½-inch pipes so that one fixture cannot use all the water the main can carry, thus robbing other fixtures. The supply lines can use T and ell fittings without disrupting the flow of water, which eliminates the need for careful measuring and fitting when you are asembling a system. As the mains course through the house they should be anchored every 3 feet or so, and be at least 6 inches apart so that the cold water cannot absorb heat from the hot water pipes. If the hot water line is wrapped in insulation, the pipes can be positioned closer to each other. Actually, if the hot water line is insulated, the water will also retain its temperature longer and reduce your water heating bills considerably.

CONSTRUCTING A WATER SUPPLY SYSTEM

The water supply pipes are small enough to fit between studs and even the 1×2 furring strips used to suspend a ceiling. Nevertheless, you will probably have to do some shallow notching in some of the joists and studs. If your pipes must run across a series of joists it might appear that the best way of supporting them is to drill a hole through each joist. That is true, but unless you are using short lengths of rigid pipe or very flexible tubing you will find it is impossible to thread the pipe through the holes. So notching becomes more practical than drilling, although by far the best way of running pipes is to hang them from the bottom of at least every other joist so that the pipes cannot vibrate and create a noise every time the water supply is shut off at a faucet.

The pipes are assembled in the basement first and then rise up through holes drilled in the floor to the first floor. They can come directly up into a first-floor bathroom or kitchen, but if you are carrying them to a second or third floor, it is easier to bring them up the same way cut through the house for the stacks. In every case, you want the pipes to go as directly as possible to their destination, not only for the sake of the system, but your pocketbook as well. Copper piping in particular is extremely expensive these days.

When you have brought the cold and hot water mains into a bathroom, install a reducer fitting such as a T or ell in the main line so that you can run ½-inch branch lines out of the walls to exactly where you want the faucet to be positioned. Use a T fitting at the point where the ½-inch branch line turns 90° to exit the walls. The center hole in the T receives a stub-out nipple and cap. The top of the T accepts a 12-inch to 18-inch length of pipe that continues up the inside of the wall and is then capped. This added piece of pipe becomes the air chamber for the faucet.

THE NEED FOR AIR CHAMBERS

Air chambers are vital components of every good water supply system; unfortunately, many older homes do have them. Water moves swiftly through the water supply pipes (remember it is under something like 60 psi), which is why it is able to gush out of a water tap when the faucet is opened. When that faucet is closed in a system without air chambers, the water comes to an abrupt halt, causing a harsh, knocking sound all through the house known as *water hammer*. More important than the noise is that the onrush of water suddenly arrested causes tremendous pressure within the system and puts an inordinate strain on every joint in the pipes, a pressure that has been known to blow both joints and the pipes themselves apart.

The air chamber stationed at each faucet provides a cushion of air for the water to push against and absorbs the shock of all that abrupt pressure. The water may even climb into the air chamber, which is why it must be capped.

When you are constructing a new water supply system, it is easy enough to add the foot or foot and a half length of capped pipe at the fitting that turns the pipe out of the wall. But if your existing system does not have air chambers, or if they are inadequate, you can purchase any of several types that can be added to a pipe after it comes out of the wall. The most popular type of air chamber is a coil of copper tubing that has one end sealed and the other end attached to a connector. But there are also rubber bags contained in a metal sleeve, and doorknob-shaped models, both of which operate on the same "cushioning" principle. The commer-

There are numerous types of add-on air chambers that can be added to your system.

cial air cushions are connected to a T fitting inserted in the supply line just prior to the shutoff valve.

If your present system rattles and knocks every time a faucet or valve is shut off, install an air cushion near each of the offending faucets or appliances, particularly your dishwasher and clothes washer. Both these appliances have solenoid valves that can turn off water so rapidly that they often cause inordinate pressure in the pipes.

Air chambers can also fill with water, which renders them useless. So if you have air chambers and the pipes still knock, close the main shutoff valve and open all the faucets in the house to drain off any water in the pipes. When no more water is coming from the faucets you can assume the pipes are empty and the air chambers have filled with air. Shut off the faucets and turn on the water again. If the water hammer still persists, try installing an extra large air chamber at the service entrance.

TESTING THE WATER SUPPLY SYSTEM

When you have run all your hot and cold water mains and brought ½-inch branch lines from them

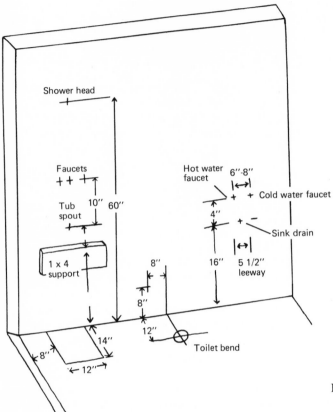

Roughing-in dimensions for a typical bathroom.

out through the walls and capped them, as well as the air chambers, test the entire system. Make certain there are no uncapped pipes anywhere, then turn on the main shutoff valve. Inspect each of the joints and connections you have made for leakage. If water appears anywhere, tighten the fitting or resolder the joint. When you can find no leaks anywhere, you may close up the walls.

INSTALLING FIXTURES

The water supply stub-out pipes at each fixture position are given a tap T with a valve at one end and a hole in the center of the T for the fixture supply line going directly to the faucet. With lavatories and many sinks, this connecting supply line will be ⅜-inch copper tubing, chromed usually. PB riser tubes also are available. Some kitchen sinks and most tubs and showers use ½-inch pipe, but it depends on the faucets you are installing. Toilets normally use a ⅜-inch fixture supply tube from the cold water shutoff valve to the flush tank assembly; they do not require hot water. They also

do not need an air cushion since the flush tank valve always closes gradually. If you are using diverters or a double faucet body with a tub or shower, these are connected directly to the branch lines and protrude from the wall to accept the faucets.

All the pipes servicing any fixture should be positioned as accurately as possible. But you do have some leeway in that the fixture supply tubes are small enough and light enough to be bent a few inches one way or the other to permit their final hookup to the fixtures. Moreover, the ⅜-inch connectors, called riser tubes, are normally flexible copper tubing. Once a fixture is in position and connected to the DWV as well as the water supply system, turn on the water and let it run down the drains to test all your connections for leaks.

REPAIRING THE WATER SUPPLY SYSTEM

Little or nothing that is very critical should ever happen to the water supply pipes or their joints

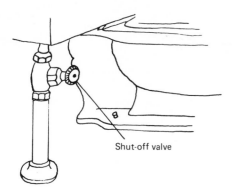

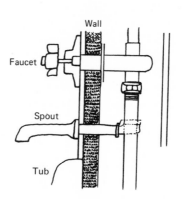

Toilets are supplied by a ⅜-inch copper tubing from the cold water main.

When walls leak, start looking for the trouble behind the nearest fixture.

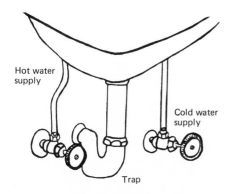

The copper tubing is flexible enough to make up for errors in measurement.

The water running through them is clean, so you should never have to run an auger through the pipes. Many years of corrosion may occasionally cause a joint to open up; either tighten the fitting or resolder it if it is copper. If that fails, replace the fitting or if the pipe has cracked, replace it.

A leak may appear in your walls. Usually, this is the result of a defective riser going to a fixture on the floor above. Always start looking for leaks at the fixture connections; the chances are that is where you will find them.

Sometimes the pipes will make noise every time a faucet is shut off. If the rattle is almost earth-shaking, you have a water hammer problem and the solution is to add air chambers at every offending faucet. If the noise is more sedate, a pipe has probably worked loose from one or two of the hangers that suspend it from a joist somewhere. Start looking in the basement, where the

mains may well run the length or width of the house. When you find a pipe that can be moved an inordinate distance back and forth, reattach it to the joists with new hangers, or reposition the old hanger. There is no reason for any metal pipe to move at all. But plastic water supply pipes—CPVC and PB—must move. In fact, CPVC pipe hangers are designed to allow the pipe to slide back and forth with thermal changes; yet hold it securely to the framing. PB's hangers need not let it slide, because its flexibility will permit sufficient movement.

The ends of water supply piping should not be locked against framing members; they must be able to move with thermal expansion. Rigid CPVC water supply piping especially needs allowance for

MAKE-UP FOR PLASTIC PIPE FITTINGS

Fitting	Center-to-End Make-Up	
	$1/2''$	$3/4''$
Socket depth	$9/16''$	$3/4''$
90 elbow, T.		
Reducing T	$3/8''$	$9/16''$
90 Street Elbow		
(streetside)	$3/8''$	$5/8''$
Coupling	$1/16''$	$1/16''$
Line stop valve	$15/16''$	$13/16''$
Union	$5/8''$	$9/16''$

(For other fittings, take actual measurements.)

expansion. (It expands about ¼ inch per 10-foot length as it becomes hot.) Straight runs of more than 35 feet should be made with 1-foot offsets (doglegs) to take up expansion. Moreover, CPVC risers that branch off from the mains should be at least 8 inches from their connection to accommodate slight thermal movement in the main. PB risers have no such requirement since they are flexible.

Support water supply piping every 32 inches, that is at every other framing member. If you are using plastic pipe, use hangers that will not cut into it as it expands and contracts.

OUTDOOR FAUCETS

You can add an outdoor garden spigot to any existing home plumbing system without great difficulty. The spigot is tapped into your cold water main at some point between the water meter and the hot water heater, but never before the water meter. If the general climate where you live has freezing temperatures, you need to buy a freeze-proof hydrant-type faucet designed to shut off the flow of water *inside* the house.

Procedure for Installing an Outdoor Faucet

1. Close the main water valve. Turn on all the faucets nearest the house entrance and drain the cold water main.

2. You can use a saddle T to tap the cold water main, but the saddle produces a reduced water flow. If you need a full flow, as you would for a sprinkler system, cut the cold water main with a hacksaw. Both ends of the pipe must be threaded so that you can install a dresser T with a ½-inch tapped opening.

3. With the dresser T in position, use whatever couplings, elbows, and pipe you need to bring the pipe as far as the wall where the faucet is to be installed.

4. Drill a 1-inch-diameter hole through the house framing. Push the outdoor faucet through the hole and connect it to the pipe with the proper couplings.

If your local code permits, the easiest material to use for this entire installation is, of course, PVC, CPVC, or PB plastic pipe, rigid or flexible.

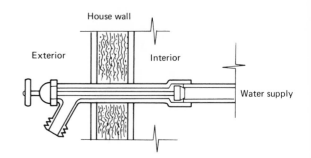

Inside an outdoor faucet.

UNDERGROUND SPRINKLER SYSTEMS

Should you decide to install an outdoor sprinkler system, you are headed back to the trenches. They are nowhere near as deep or wide as a sewer disposal field, but you are still in for a considerable investment of time and money.

Water is the essential ingredient to any good lawn and the kinds of automatic sprinkler systems that can keep your lawn in class A condition were pioneered by the golf courses, whose business it is to have well-manicured grass. For years, farmers have given their crops consistent and proper amounts of water by using versions of these same sprinkler systems. The systems consist of a series of underground pipes that bring water to sprinklers at preestablished times. The water flow is controlled by an electronic control unit that, for your purposes, costs in the neighborhood of $100. The control unit opens and closes valves leading to different sections of your lawn at different times. The lawn is divided into sections because with a normal entrance pipe of ¾ inch or 1 inch your water service could not effectively sprinkle more than one area at a time without losing considerable pressure.

The sprinkler system you install can be as simple or as complicated as you wish. The most basic system consists of an outdoor faucet connected to an underground pipe having six or seven pop-up sprinkler heads attached to it via T fittings. From that kind of simplicity you can have pumps that will pressure-feed water through a dozen or more sprinklers, or a clock-operated programmer that turns sections of the system on or off through electronic or hydraulic valves. How much anyone spends on a system will vary from lawn to lawn but a rule-of-thumb estimate is $.15 per square foot of lawn, which includes the pipe, control valves, heads, fittings, and controller.

The best way to design a system that suits your particular lawn is talk to local sprinkler system dealers. They can give you specific information concerning the valves, sprinkler heads, and control units that will allow you to evaluate the kind of equipment you need and its cost.

In any case, the sprinkler system must include a vacuum-breaker valve to prevent any cross-connection between the underground sprinklers and your potable house water supply. The sprinklers, of course, are exposed to weed killers and fertilizers, none of which you would want in your family's drinking water. The vacuum-breaker valve provides an automatic air gap to keep that cross-over from ever occurring.

The pipe used in a sprinkling system is either PVC or PE plastic pipe with a nominal diameter of ¾ inch. The pipes lead away from a manifold that contains the valves and is buried about 6 inches under the surface of the ground. Each pipe feeds several sprinkler heads.

Fortunately, you do not have to churn up half your lawn to lay the pipes. You can use a small trencher, which will cut a slot about 1 inch wide and as deep as you need. The narrow strips of fresh earth left by the trencher will disappear under new grass within a few weeks. An alternative way of digging the trenches is to make a V-shaped slot in the ground with your spade. Lift each wedge out of the ground and tuck your pipe in the bottom of the V, then force the sod back down over the pipe. The scar in your lawn will vanish almost immediately.

The number of heads attached to a given sprinkler line depends on the type of head you use, the pipe size, and the amount of water pressure delivered to the pipe. You must have at least 40 psi and all of the heads available can be adjusted to spread various size circles of water, so again you will have to consult with your dealer to decide on the best possible combinations of pipe size, heads, and control unit.

The control units offer as many variations as sprinkler heads, and obviously, you do not want to pay for a unit that has a lot more capabilities than you will ever need. All the pipes you lay will ultimately come together near your house and are connected to the service entrance between where the main pipe comes into the house and the meter.

The manifold is a series of valves that should be protected by an enclosure of some kind built around them. The control unit is usually placed inside the house. It is not only connected to the valves but must also be plugged into an electrical outlet.

SOURCES OF WATER

The moment you locate a house beyond the range of a municipal water supply system you are confronted with the problem of where and how to get the fresh water needed for everyday living. Rural and vacation homes can be supplied in any of several ways, most of which reduce themselves to some form of well that draws fresh water from an underground spring or water table. You can also take your fresh water from a surface spring, lake, river, or pond; but you must assume that any surface water, including pure rainfall, can be contaminated by humans, animals, and the atmosphere. So in most cases it must be purified, usually by adding chlorine to it, before it is safe enough to drink.

SURFACE WATER

A constantly wet spot in the earth, a stream, a lake, a pond, can all be tapped as a surface water supply, but the water must be collected, contained, and then controlled without exposing it to pollution. It may flow over rocks or sand, but not soil because dirt contains all kinds of living organisms that will make any water fail all the purity tests. How much water you tap per day depends on how large your family is, but you will probably need to have at your command something like 500 gallons a day, figuring between 60 and 70 gallons per person per day, plus a reserve.

Tapping Surface Water

If your source of surface water is on a hillside somewhere above the house, you can tap the water by digging a trench from it to a collection tank. Lay a perforated pipe in gravel and back-fill the trench; the pipe should be sloped and connect to a watertight plastic pipe that leads to an enclosed collection tank. The tank or cistern can be made of

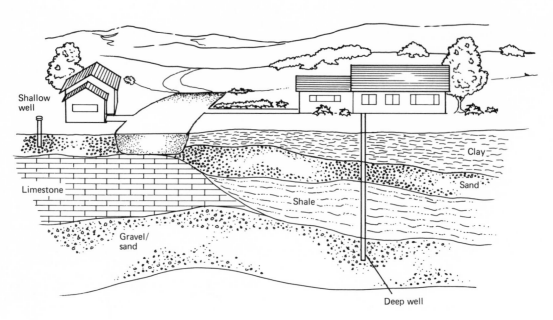

Where the water can be found.

either steel or concrete and should be large enough to hold a full day's supply of water.

You might also tunnel back into a hillside to your water source and then build a series of small dams, which will collect most of the silt in the water before it enters your collection pipe. You are always trying to collect the purest, least silted water and anywhere you can tap into a surface source at a point where there are rocks or sand, do it. Also do it at a place as high above the house as possible. If the water must drop 100 feet to your house it will build up 45 psi of pressure all by itself and you can flow the water from your cistern to the house without needing a pump to build up pressure. If, however, the spring is at only a slightly higher elevation than the building, or if it is below the house (or if you are using a well), you will need a pump.

WELLS

There are driven, dug, and drilled wells and most of them should be made by professionals having the proper skill and equipment. As soon as you decide that you need a well dug, you must also answer the obvious question, "Where?" The expertise of a professional can be invaluable in locating your well.

The well should be placed as conveniently to the house as possible, but having selected where you would prefer to have it, you must also find out if there is any water under the site. You can turn to the science of geology and it will tell you if there is water somewhere in the ground below your chosen site. You can also use the not so precise, but often as accurate art of dowsing, or water-witching. The geologists will make a studied survey of your property and present you with a series of calculated estimates of where the water ought to be. The water dowser will hold his (or her) forked tree branch in front of him and walk around your property until the branch dips toward the ground. You cannot do that successfully yourself, unless you happen to have the talent for it. It is, admittedly, a stretch of your blind faith to spend your hard-earned money drilling for water with nothing but a piece of wood as verification. But water-witching works. It has worked for centuries all over the world; just be sure to hire a dowser with a good reputation for finding water.

Driven wells are the simplest kind of well to dig. They are made by hammering a pointed pipe into the soil with a sledge hammer until you reach water. The pipe has a screen and perforations in its pointed head, which allows water to enter and fill it as soon as it reaches the water table. The screen filters out much of the sediment and the water can be brought to the surface with a suction pump lowered down the pipe. Driven wells are inexpensive and easy to install, but unfortunately they cannot go very deep into the ground and the pipe cannot be hammered through rocks. They are therefore used mostly for collecting shallow ground water in places that are primarily made up of coarse sand.

A dug well can be made by hand or with power tools. Usually, a dug well is 3 or 4 feet in diameter and must go deep into the water table to be effective. The sides of the well are kept from crumbling by inserting concrete well rings. But for all of this, dug wells are rarely more than 50 feet deep.

By far the surest method of getting a well is to drill it. You need some formidable hardware to drill down through earth and rock formations and to keep drilling until an adequate supply of pure water can be tapped. Which is why drilling wells can hardly be classified as a do-it-yourself project and why you will have to hire a reputable drilling contractor. Drilled wells can go either straight down, or horizontally into a hillside; and when the water is reached, 4- to 6-inch-diameter casing pipes are installed in the well hole and capped to prevent being contaminated by near surface or aboveground pollutants. You will probably need a pump to bring the water from its source to your house.

WATER PUMPS

There are different pumps for drawing water from shallow, medium, or deep wells; and the capacity of the unit you select should depend on the depth of your well and how much water you require.

Shallow-well, piston-type pumps will handle water depths of up to 22 feet and can be either hand-operated or automatic. The simplest of these consist of a cylinder, a piston, and two valves. When the piston rises, it opens a check valve drawing first air and then water into the cylinder. The piston then descends, closing the check valve and at the same time opening a valve at the top of the cylinder so that the water can pass out of the pump. The water, once the pump has reached its operating speed, is able to flow continuously from the well to the house.

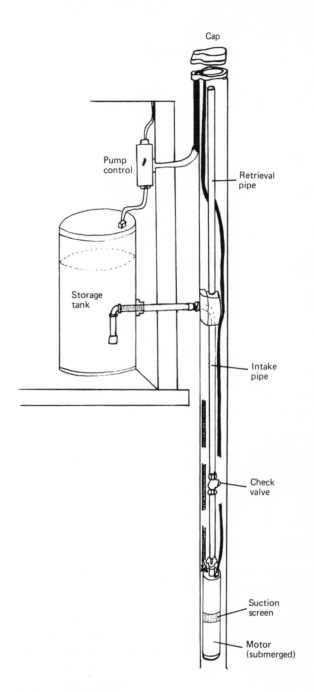

Submersion-type pump.

The automatic version of the piston-type pump has four valves and a double-action piston that draws water into the pump and pushes it out on both the upward and downward strokes of the piston, producing a steadier flow of water.

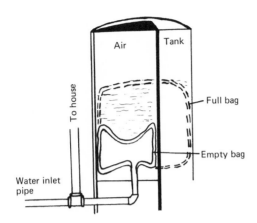

How the sealed air tank works.

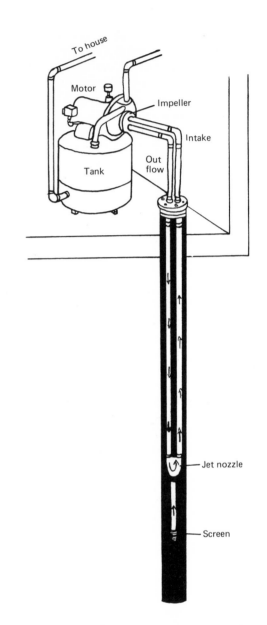

How a jet pump works.

Centrifugal pumps are also used with shallow wells, but these must be primed before they will start. Rather than valves, the pump has an impeller, or multi-vaned fan, positioned in the discharge port of the pump. When the impeller spins, it sucks water out of the ground and pushes it into your collection tank. The reason the pump must be primed is that all the air in the pipe must first be got rid of, or all the impeller will do is pump air. So to prime the pump, you must fill the pipe with water.

Medium and deep wells require either of two different kinds of pumps: the jet-type or the submersible. The submersible pumps are used in sandfree wells up to 500 feet deep and although you control them from inside your house, the pump and its attendant electrical wires are actually positioned at the bottom of the well. The submersibles are compact and buried so deep in the ground that you cannot hear them or feel any of their vibrations. The pump brings well water into a sealed air tank, which compresses the air in the top of the tank until the pressure reaches the desired 40 or 50 psi. At that point a pressure switch shuts off the pump and a check valve prevents the water from running back into the well. As the water is used, the pressure diminishes and the switch automatically turns on the pump again.

The centrifugal, jet-type pumps are less expensive than the submersibles and many experts recommend them above any other pump. The

motor and its turbine pump are situated above the well and a suction pipe is extended down into the water, along with a smaller jet or pressure pipe. Water is then pumped down the pressure pipe and bent around and back up through the jet, then is carried on upward through the suction pipe to the pressure tank in your house. As the water reaches the pump at ground level, some of it is diverted back to the intake side of the pump to keep the flow of water moving; most of the water goes into a pressure tank, which holds it for use in the house.

The delivery of water by any pump depends on how far the water must be lifted, and how much pressure it must be under when it gets to your house. The higher the pressure needed, the fewer gallons per hour the pump will be able to deliver. Computed at between 60 and 70 gallons per person, per day, whatever kind of pump you buy should be able to give you a minimum of 300 gallons per hour. It is preferable if it can provide 550 gallons.

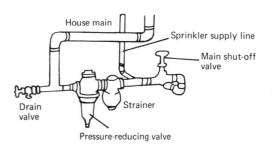

A pressure-reducing valve and its strainer.

REDUCING WATER PRESSURE

If your house service enters the building at much less than 40 psi, you need a pump to get the water pressure up to snuff. But the reverse situation can occur as well, where you have so much pressure in the pipes that faucets leak or wear out quickly; there are all kinds of noises going on throughout the system; and considerable water is just plain wasted at the fixtures, to say nothing of the very real increased danger of water hammer blowing apart the whole system.

The water pressure in your system should not exceed 60 pounds per square inch; the way you determine that pressure is to connect a pressure gauge to any hose faucet. The gauge, by the way, will cost you less to purchase than if you call in a plumber to do the testing for you.

If your water pressure is too high, you should install a pressure-reducing valve. The valve is attached to the main entrance line just past the main shutoff valve and comes with a union for easy hookup. Even so you may have to cut out a piece of the main pipe so the unit can fit in the line. Most pressure-reducer valves are set by the manufacturer for 50 psi, but there is an adjustment knob on the bottom of the unit so that you can alter the pressure if you wish.

SOFTENING AND PURIFYING WATER

The purest form of water is rain, but as it filters down through the atmosphere and then the earth, it absorbs such minerals as calcium and magnesium, some of which it absorbs and some of which is likely to remain suspended in the water. When these materials remain undissolved the water is known as "hard"; most of the water found in North America and Canada is rated between 3 and 30 grains per gallon of calcium carbonate, which ranks it as between "moderately" and "extremely hard" water.

Hard water needs more detergents, leaves rings in your bathtub, and streaks on your dishes. It makes your laundry gray, clogs your pipes, and makes cooked vegetables tough. You can also get a reddish color to your water from colloidal iron, black deposits from manganese, or the water may smell like rotten eggs, which comes from too much hydrogen sulfide. Corrosive water contains acids and oxygen, and turbidity is caused by an overabundance of fluorides and nitrates.

Most of the problems that arise with hard water can be solved by adding the proper equipment to your water supply system and most of these units are easy enough to install without needing professional help. The question, of course, is what problem do you have and which units do you install? You can test water for hardness by pouring distilled water in one glass and tap water in another. Now add ten drops of liquid dishwashing detergent to each glass and shake the glasses until you have made them both sudsy. Compare the suds level in the glasses; the water is hard if the glass with tap water has less suds than the distilled glass. If there is a difference in the suds level, add ten more drops of detergent to the tap water and shake it some more. If the suds levels are identical, at this point you can assume your water is twice as hard as pure water. If the tap water is still not equal in suds to the distilled water, add another ten drops of detergent and shake the glass again. Continue adding detergent until the suds in both glasses are equalized. If it takes fifty drops of detergent in the

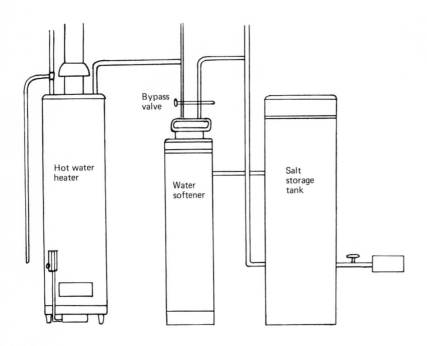

Placement and hook-up of a water softener and its storage container.

tap water to equalize the detergent in the distilled water, your tap water is five times as hard, and so on.

Now that you know how hard your water is, you still may have to ask your local equipment supplier to test it for you (in most cases at no cost) to determine the kind of hardware you need to solve the problem. The most usual water treatment is to simply soften the water, which is accomplished by running it through a mineral or synthetic resin compound. Chemically, this exchanges the "hard" magnesium and calcium ions, which are electrically charged particles, for "soft" sodium ions.

Water softeners can be rented on a service contract or purchased outright; the difference is that if you own the unit you must regenerate it yourself, which amounts to pouring salt into it from time to time. Some units have automatic sensors that regenerate it whenever the salt content is low; others are semiautomatic and require you to add the salt physically. Because of their gallonage of salty effluent produced during regeneration, do-it-yourself softeners are beginning to be curtailed by some authorities, in favor of factory-regenerated softeners.

There are also numerous types of water-softener appliances designed to handle such problems as removing sediment, hardness, or a high iron content. You can find them in appliance-styled cabinets that are suitable for installation in your kitchen, or in tank-type versions that are placed in your basement or some other out-of-the-way area. If you have the space, any of the tank models will provide more storage of both water and salt and are therefore more convenient. But no matter what you buy, make sure it has the seal of approval from the Water Conditioning Foundation, which in the water-softening business carries the same respect as the Underwriter's Laboratories does with electrical equipment and supplies.

Only some of the water used by a household actually needs to be softened. Hose water, used to water your grass or plants, and normally the water used in toilets can be hard, particularly if the flush valve is made of space-age plastic rather than metal; in many parts of the country only the hot water is softened, When it comes to installing a water-softening unit you have a variety of options as to how much water you intend, or really need, to soften. The biggest decision involved with installing a water softener is where to hook it into your system. If you plan to soften only the hot water, the softener can be connected to the cold water line just before it feeds into the water heater. If all water in the house is to be softened, the softener must be attached to the main service pipe at a point beyond the water meter, main shutoff valve, and the branch lines to your outdoor faucets and sprinkler system.

The softeners get rid of water hardness through the process of ion exchange and they do this by forcing the water through a storage tank full of softener salt pellets or rock salt, which gives up its sodium ions to make the water soft. Later, when all the sodium ions have been exchanged, the mineral bed is regenerated by passing salty water slowly through it. This comes from softener salt pellets or rock salt through which the water has passed. The sodium in salt passes into the mineral bed until it has all it can hold.

When installing a water softener, use flexible piping, which is connected to the water main with elbows and the appropriate adapters. Cut about 8 inches out of the main line and install an elbow on each of the open ends, then bring the flexible pipe to your inlet and outlet openings in the softener unit and connect them. (Exactly how you connect them depends on the unit; see its accompanying manual.)

There is a safer, or at least more convenient, method of connecting the softener that allows you to continue having water service during whatever time the softener is removed from the system to be regenerated: Cut 8 inches of pipe out of the main line and install Ts at both open ends. Connect a gate valve on the main line between the Ts. Also install shutoff valves on the hard water inlet and the soft water outlet lines going to and from the softener. Most of the time the gate valve will be closed and the shutoff valves open. But should the softener have to be disconnected, you can close the shutoff valves and open the gate valve and have uninterrupted water service (even if it is hard) while the softener is being regenerated, that is, having its salt replaced.

Waste water from the softener unit can be released directly into the ground, or be connected to the sewer via a trap. It will not affect a septic tank other than the fact that it is extra water that is probably better kept out of your disposal field if possible.

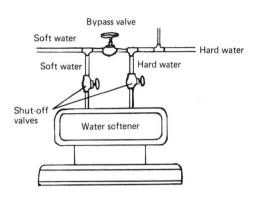

The best way of connecting a water softener is to use a bypass valve in the water main.

Removing Impurities

When you have identified the impurities in your water system you can choose one or more impurity-removing units to improve your water. All the water softeners and filters are easily installed; the feeders are more complicated and you might find it more practical to have them installed and serviced by professionals.

High Iron Content

Water softeners can—and do—remove iron from the water, but if you live in an area where the iron content is unusually high you will need an iron filter which can remove and/or dissolve most of the suspended iron in the water. The filter contains minerals that trap the iron particles, so it must be flushed out periodically; if there is a high oxygen content in your water the filter may have to be flushed with potassium permanganate, which is poisonous. The filter then has to be thoroughly flushed out before it is used again.

Sulfur Water

Water containing sulfides can be cleared by using an iron filter together with an activated-carbon filter. The activated-carbon filter is typically a canister connected to the cold water line under the kitchen sink along with a shutoff valve. The unit has a pressure-relief valve and a glass canister that threads into the unit so that it can be easily removed and the filtering element replaced.

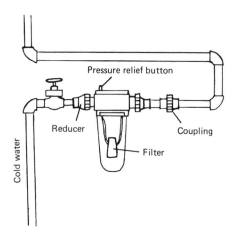

An activated carbon filter.

Mildly Contaminated Water

Automatic chlorine feeders are designed to inject chlorine into the water supply to eliminate bacteria that may be slightly contaminating the water. But the chlorine also creates some undesirable particles and usually the feeder must be used in conjunction with a sediment filter. The particular combination of units will do a good job of removing iron from the water, while a chlorine feeder used with an activated-carbon filter can improve water that contains too much sulfur.

Corrosive Water

Phosphate feeders inject phosphate crystals into the water main where they dissolve slowly into the water. The neutralizing filters also add chemical agents to the water to reduce its corrosiveness. Both units need to be replenished from time to time, but if your water is inordinately corrosive, the units will coat the insides of your pipes (and the hot-water heater tank) and slow the process of corrosion considerably.

Cloudy Water

A water softener will often clear up cloudy water. Perhaps more reliable a solution is to use a sediment filter plus an activated-carbon filter. The sediment filters often contain sand or gravel, both of which will catch much of the particles clouding

the water, but if the clouding is extreme, try a chemical feeder instead of the activated-carbon filter.

Bad-Tasting Water

If the water tastes lousy, an activated-carbon filter may be enough to improve it. But if the water is also brackish or salty, or if it must be delivered for drinking free of sodium or other minerals you will have to use either a reverse osmosis or a deionization water treatment.

Reverse Osmosis Units

Reverse osmosis units can cost in excess of $200 and have a membrane in them that must be replaced every two years at a price of about $40. The unit is usually connected to the cold water line under your kitchen sink; it has no moving parts and does not have to be regenerated. It operates solely from the water pressure in your household system. The membrane inside the unit is constructed of cellulose acetate wound around a carbon filter, which separates the impurities from the water. The water that passes through the unit is pure enough to drink and cook with, as well as to use in both steam irons and car batteries.

Water Deionizers

Deionizers are sometimes called demineralizers and are invaluable under certain water conditions, but they are also costly to operate. One version is designed for connection to the plumbing, but you can also get one that attaches to a faucet, which is generally less expensive to replace than to regenerate when it loses its effectiveness.

Deionizers will remove nearly all dissolved solids in your water and render it pure enough for most household uses. To do their job they go through some tricky chemical processes that rely on the use of two ion-exchange resins. The first mineral used is the same one found in water softeners except that it is charged with hydrogen ions. instead of sodium ions. As water passes through the mineral, the positive hydrogen ions are released in the water and hold back the positive minerals, such as carbonates and iron. The water then passes through a second bed of resin that is charged with negative hydroxide ions. These are

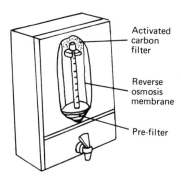

Activated
carbon
filter

Reverse
osmosis
membrane

Pre-filter

The reverse-osmosis unit.

released in the water to hold back any negative mineral ions, such as chlorides and bicarbonates. Then the hydrogen and hydroxide ions are free to combine in the water and purify it. All these ions get tired after about two years of daily use and must be regenerated by using a strong acid in the hydrogen resin and an equally powerful alkali in the hydroxide. The process of regeneration normally requires that the unit be removed from the plumbing system and taken to a chemical plant. At best, the regeneration should be performed by a professional.

8 Hooking Up Appliances

There are three major appliances used in the home that require plumbing as part of their installation, as well as electricity. All three, the clothes washer, dishwasher, and garbage disposer, are relatively easy to connect into your house plumbing system, although with the clothes washer, at least, you may have to open some walls or part of your floor.

THE GARBAGE DISPOSER

As an appliance, the garbage disposer offers several advantages, beginning with sanitation. There are two kinds of disposers, continuous and batch feed, and which one you purchase is a matter of personal preference, rather than performance. The continuous feed machines have an open drain port, which you can stuff garbage into as long as the machine is running and it will continue grinding. The batch feed machines are filled with waste and then activated when you lock the drain lid in place. If you have more waste than the machine can hold, you must wait until it has digested whatever you put in it and then undo the lid and fill the unit again.

All waste disposers are canisters suspended under the drain in a kitchen sink and their drain is connected via a trap to the kitchen sink drainpipe. The canister has a powerful capacitor-start motor in its bottom, which is sealed from the top portion of the tank. The grinding chamber contains a "cheese grater" around its inside called a shredding ring and a rotating disk spins inside the shredder ring. The flywheel contains two or more flyweights that throw the waste against the shredder ring as the wheel rotates at a high speed. The machine snarls and groans and coughs and gurgles as it grinds waste into minute particles, which are then washed out of the drain port and down the sewer. Disposers will grind up practically anything except paper, metal, or large bones. They will also grind up your best silverware, so you have to be careful not to let any family heirlooms fall into the machine—the silverware will be a mess, and the machine will most likely need its shredder ring and flyweights replaced.

Installing a Waste Disposer

You will probably spend more time wrestling with the electrical connections of a disposer than you will with the plumbing. All units come with complete installation instructions, but it helps if you have a working knowledge of house wiring procedures. Doing your electrical wiring around well-grounded plumbing pipes and fixtures can get you killed. So be careful and follow all the standard safety procedures suggested by the National Electrical Code. At least a #14 gauge nonmetallic

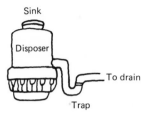

A garbage disposer is large enough to fill much of the space under a kitchen sink and must be connected to the sink drain.

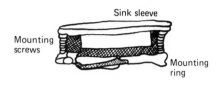

The disposer mounting consists of a mounting ring and a sleeve, which replaces the sink strainer and is held to the underside of the sink by mounting bolts.

sheathed, or BX cable must be run from the motor in the bottom of the disposer to a standard toggle switch such as is used with any wall light. Then the cable is connected to an existing outlet or junction box or run directly to your fuse box. Neither the switch nor electrical cable is sold with the appliance but must be purchased separately.

While you are buying electrical supplies for the disposer, you may also have to purchase some sink drain hardware, called tubular goods, for the main hookup, including a new trap. Anywhere the code permits, the easiest drain connections to make are with ABS or PVC plastic tubular goods. Polypropylene has the advantage of being the most chemical-resistant of the three. It will take anything you would be likely to pour down a drain. All can be purchased as a drain kit (including a trap, connecting drainpipes, and slip nuts) that allows you to assemble a variety of configuations under your kitchen sink. The versatility of the drain kit is useful because you may have one or two sinks and you may or may not want to connect your dishwasher directly into the disposer. Most units have a special knockout in the side of the canister so that a dishwasher can be connected into them. The connection is there not because your dishwasher needs the disposer to grind anything up for it; it is because the plumbing arrangement under your sink may be such that the only way you can have both appliances is to interconnect them.

Installing a Disposer

While the exact plumbing assembly will vary from unit to unit, the procedure for installation is generally the same for all disposers:

1. Remove the existing sink trap, strainer, and drain line. The trap and drain line come apart by

rotating the couplings and do not be fooled by the strainer. It resides in an unthreaded hole cut in the bottom of the sink, but has a locking nut threaded around its base which must be unscrewed. The strainer was installed in a bed of putty, which may have glued the unit to the bottom of the sink. You may have to pry it, twist it, or bang the bottom of it with a hammer to get it loose. Whatever you do, be assured that it is not threaded into the sink itself, even though it may be stubborn enough to make you think it is.

2. When the old drain assembly is removed, scour the area around the drain hole on both sides of the sink bottom until the porcelain or metal is absolutely clean.

3. Remove the entire sink mounting assembly from the top of the disposer. The assembly will include a snap ring, mounting screws, a mounting ring, a fiber gasket, and a sink sleeve. Dismantle the assembly by loosening the mounting screws until you can pry off the snap ring.

4. Apply a ring of oil-base plumber's putty around the top of the drain opening at the bottom of the sink and place the sink sleeve in the opening. Push down on the sleeve; an ample amount of putty should squeeze out from under the sleeve, but do not remove the excess yet.

5. From under the sink, slide the fiber gasket up around the sleeve and tighten the back-up ring against the bottom of the sink. Slide the mounting ring over the sink sleeve and insert the snap ring into its groove on the sink sleeve to hold them in place. Tighten the mounting screws in the mounting ring evenly and as tightly as you can get them to hold the mounting ring securely in place against the bottom of the sink. You will notice that the mounting ring has angled flanges around its bot-

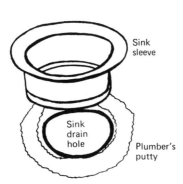

The sink sleeve is embedded in a ring of putty around the sink drain hole.

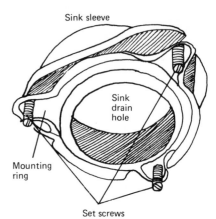

The sleeve is threaded at its bottom, allowing the mounting ring to be screwed up under the sink drain hole.

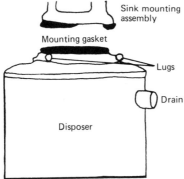

The disposer has three mounting lugs on its neck that must fit into the angled slots of the mounting sleeve.

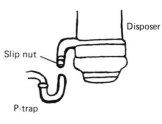

With the disposer hanging from the underside of the sink, install the unit's drainpipe and a P-trap. The trap is connected to the sink drain line.

tom rim. These correspond to mounting lugs attached to the top of the canister, which will slide under the flanges and hold the unit in place.

6. If you intend to connect your dishwasher to the disposer, you must first remove the steel and rubber knockout plugs in the side of the canister to form the disposer dishwasher inlet. You do this by placing a blunt instrument against the plug at an angle and hitting it with the heel of your hand. The plug or plugs will fall inside the unit and you'll have to reach in and retrieve them. Most manufacturers sell a dishwasher installation kit for their disposers so that you can make your connections properly and with the least amount of effort.

7. Prepare the disposer for accepting the dishwasher drain line before you install the canister under your sink. The drain outlet in the side of the canister must be fitted with an elbow or straight piece of pipe, which is usually supplied by the manufacturer. The pipe is flanged at one end. A gasket is placed over the flange and resides between the pipe and the drain outlet on the side of the disposer. The gasket and pipe are held in place by bolting a metal plate around the flange to the canister.

8. When your drain connections have been installed, lift the disposer up to the mounting assembly and rotate it into the mounting ring until the lugs are tight. You will most likely discover this is not as easily done as said. There are usually three mounting lugs and they have a way of catching two of the mounting flanges but not the third, which means you have to keep holding that heavy unit up

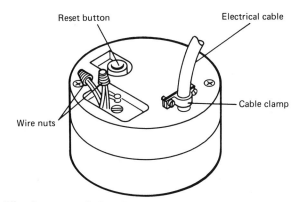

The bottom of the disposer contains a motor reset button and positions for inserting an electric cable from the switch. The wire connections are made in a well that is sealed from the motor and cannot be damaged by water.

under the dark sink and feel your way behind the ring to make certain all three lugs are properly engaged. You must also get the unit tight against the sink but turned to exactly the direction you want the drain lines to go. The process can take a while to accomplish.

9. Now connect the discharge tube on the side of the disposer to the sink drain. Unless you have brand new plumbing it is a good idea to run an auger through the drain line to make certain it is in no way clogged. The drainpipe you attached to the disposer is, most likely, plastic but by using a slip-nut connection on the trap you can also connect your existing metal drain trap. However, the canister takes up space under the sink and you will probably find you have to redesign the trap configuration to make it fit properly into your drainpipe. It's easiest to start with all new tubular goods.

The electrical work involved with a disposer installation can be done either before or after the unit is hanging under the sink. There is a well in the bottom of the unit, which is covered by a plate that must be removed. About the only thing you will find in the well are black and white lead wires from the motor and a grounding screw. Connect your cable to the wires with wire nuts and the bare grounding wire to the grounding screw, then put the cover back on the well. So far as the disposer is concerned, that is all the wiring you have to do. The remainder of your electrical work involves finding a place for the switch and a junction or outlet box that you can connect the cable into.

DISHWASHERS

Automatic dishwashers divide into either top loaders or front loaders; the top loaders are almost all portable and the front loaders are permanently connected to their electrical and plumbing connections. With the portable units, which are rolled around on casters, both the water supply and drainage hose are attached to the sink and then put back in the appliance when it is stored between uses. Installation of the permanent models can vary considerably, so whatever model you are putting in your house, follow the manufacturer's installation instructions scrupulously. In general, bear in mind the following suggestions when putting in a dishwasher:

1. Be certain the water pressure is between 15 and 120 pounds per square inch (it is probably between 40 and 60 psi, but make sure). If the dishwasher receives less than 15 psi, it will not fill properly. If the water is coming in at more than 120 psi, install a pressure-reducing valve ahead of the machine's water inlet valve.

2. Make sure the machine will receive sufficient amounts of hot water at its required temperature. In most cases this means at least 140°F. All dishwashers have some form of heating element in them but its function is not to bring the water up to 140°F; it is there to maintain water temperature and to dry the dishes.

3. All your plumbing should be done in strict accordance with local plumbing rules. There will be local restrictions on traps, vents, and the kind of pipes you can use. Some codes require air gap devices to prevent a cross-connection. In most cases the water supply lines must be ⅜-inch copper tubing. The drain should be ½-inch copper or plastic tubing. It is a good idea to include a shutoff valve in front of the water inlet valve to facilitate any servicing that may have to be done during the lifetime of the appliance.

4. Permanently installed dishwashers are normally wired directly into the house wiring system and must conform to the National Electric Code as well as any local codes. The power line must be able to supply the pump motor and heating element without a noticeable drop in voltage, so the minimum gauge wire you should use is #14 with 5-ampere fusing. A #12-gauge wire with a 20-ampere fuse is mandatory, if your garbage dis-

poser is supplied by the same electrical circuit as the dishwasher. It goes without saying that all code rules pertaining to proper grounding should be observed. The manufacturer's installation manuals are also very specific about grounding their machines. Follow their instructions closely.

The dishwasher can, of course, have its own source of water coming from the basement, and it can drain directly into the main stack, just as it can have its own power line. But dishwashers are usually installed close to the kitchen sink and it is under the sink that offers the easiest plumbing connections. You can tap the hot water line as soon as it comes out of the wall by inserting a T fitting between the pipe and the shutoff valve that will accept the ⅜-inch copper tubing coming from the dishwasher.

The same thing can be done with the kitchen drain. Simply insert a T somewhere in the drainpipe that can carry the ½-inch dishwasher drain. Some tubular goods manufacturers offer "dishwasher Ts" that can be inserted ahead of the trap in the sink's drain lines. These include a side tapping for connecting the dishwasher drain. Otherwise, the drain line to the dishwasher should include a trap and, as a sanitary precaution, an air-gap device that will prevent any sink waste from backing up into the dishwasher. When installing the check valve, make certain the arrow printed on its housing is pointing *away* from the dishwasher.

WASHING MACHINES

Washing machines are installed in basements, kitchens, bathrooms, and special laundry rooms. They can be upstairs, or downstairs, or all around the house—so long as they are not too far away from the house DWV and supply lines. The plumbing connections can be made under the floor, in the walls, or any other place you desire. Washing machines almost always come with a three-prong plug that must be inserted in a three-hole grounding receptacle. Nobody ever recommends you use any kind of an extension cord with a washing machine. The washer should have its own 120-volt, 15-ampere circuit.

The plumbing that attends a washing machine is complete and you must follow all of the rules; the supply lines require extra large (¾ inch by 18 inches) air cushions and standard shutoff valves, just like any sink; the 1½-inch drain does not need a trap, but it should have a stand pipe as well as be vented. The good news is that clothes washers arrive from the manufacturer with all the appropriate hose connections, so all of your plumbing work is concentrated on the house system, not the appliance.

Tap into your hot and cold water supply lines at a point as close as you can get to where the machine will be standing. Your lines should not have to run more than 5 feet. Both lines should have 12-inch to 18-inch capped air cushions and end with threaded hose-bibb faucets that will accept the hot and cold water inlet hoses attached to the machine.

Washing machines must receive their water supply under a pressure of between 20 psi and 120 psi. More than 120 psi will damage the mixing valve in the machine so if you have more than that you will have to install a house pressure-reducing valve. If the water pressure is below 20 psi, the machine will have an abnormally long fill time and poor spray-rinse action. It may also develop some leaking water valves since water pressure is the only way the valves can close properly.

Also consider the hot water. It must be at least 140°F (preferably 160°F) when it reaches the washer, and some washer cycles may use as much as 30 gallons of hot water, which is the total storage capacity of smaller water heaters. In order to operate a clothes washer, the hot water should be supported by at least a 50-gallon, quick-recovery electric water heater or, at the absolute minimum, a 30-gallon, quick-recovery gas heater.

Even if your heater is large enough to supply the washer, there is the problem of delivering 140° water into the machine. You may have to set the water temperature in your heater above the 140°F "warm" level in order to obtain the recommended 140°F–160°F temperature for the washer. An alternative is to insulate the exposed hot water supply line between the heater and the washer. An unprotected hot water line can lose as much as 1°F for every foot of its run. If your hot water is set at 140°F and the clothes washer is 50 feet away from the heater, it is receiving water that is only 90°F, and to get it up to the 140°F minimum requirement, you would have to set the water heater thermostat at 190°F, which is higher than heaters are gauged to supply. Even at 180°F, the highest

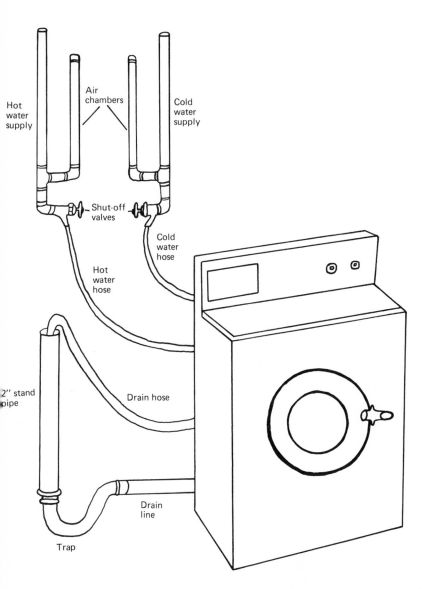

Hot
water
supply

Air
chambers

Cold
water
supply

Shut-off
valves

Cold
water
hose

Hot
water
hose

2″ stand
pipe

Drain hose

Drain
line

Trap

Anatomy of a clothes washer water supply and drainage hook-up.

"hot" setting on water heaters, you may have excessively hot temperatures in your kitchen and bathrooms, where small children (and adults for that
matter) may inadvertently be scalded. Wrapping
insulation around your hot water pipes wherever
they are exposed will prevent almost all heat loss in
the hot water pipe and may be the most economical
solution to getting properly heated water into your
clothes washer.

The waste water from a clothes washer may be
removed from the machine via a laundry tub, stand
pipe, or floor drain. If you are using a laundry tub,
make sure it has a larger capacity than the washer—
obviously if you have an 18-pound washer
that uses 25 gallons during its washing cycle, a
20-gallon laundry tub is not a large enough drain
receptacle. The tub or drain must also be situated
so that it provides proper drainage for the washer

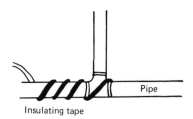

Insulating tape

The hot water pipe between your hot water heater and the clothes washer or a dishwasher should be covered with insulating tape, to reduce heat loss.

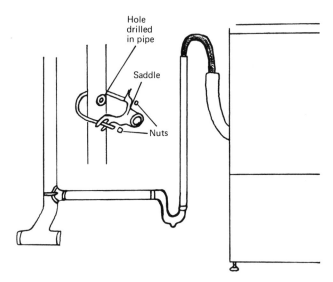

Hole drilled in pipe

Saddle

Nuts

The standpipe can be connected to your drain line with a saddle, which is easier to install than it is to cut the drain line apart and insert a T.

drain hose. In general, washer drain hoses must discharge at a minimum height of between 30 and 34 inches and no more than 72 inches. If the height from the base of the washer is less than 30–34 inches, water may be siphoned out of the machine during its wash-and-rinse cycles; if it is more than 72 inches, the unit's pump may not be able to get all of the water out of the washtub.

The drain hose supplied with the machine is the maximum recommended length of the drain, so do not add more hose to it or you are liable to run into "sagging" or kinking problems; that will result in siphoning water out of the machine when it is not supposed to be drained, or kinks that restrict the draining altogether.

If you are using a stand pipe, it must connect directly into your DWV system, slope at ⅛-inch per running foot, and should be a minimum of 2 inches in diameter. It comes out of the floor and stands vertically between 30 and 34 inches (and not more than 72 inches) above the floor. The drain under your stand pipe must be trapped in accordance with local plumbing laws and have at least a carry-away capacity of 18 gallons per minute to guarantee proper draining and prevent any overflow of water onto the floor.

9 Water Heaters

As soon as cold water enters your home it passes the main shutoff valve, then the water meter and starts its run through the building to service all of your fixtures and appliances. The first of these appliances is your hot water heating device. The water that is diverted from the cold water main to the heater is warmed, then pushed out of the unit and into the hot water main, which you will almost always find running throughout the house parallel to the cold water line. Almost without exception, the hot water line is connected to the left side of each fixture, and has an air cushion attached somewhere near the faucet.

The device that heats all of the water in your home can be one of several types, from a set of pipes or coils, which reside in a separate chamber at the back of your furnace, to gas, oil, solar, or electrically powered heating units. No matter how it is fired, the most important fact about any heater is its size. The proper sized heater for a specific household is determined by how many people it must regularly serve; how much water those people will use during peak times of the day when everyone is bathing or laundering; and the designated recovery rate of the heater itself. Generally, you can assume that your heater will have to deliver 10 gallons of hot water per person, per hour, during peak use periods. A family of four, then, needs at least a 40 gallon heater. That does not mean the heater must always operate at capacity,

but it should be able to generate enough hot water to meet the 10 gallons per person per hour estimate.

The recovery rate of every heater is determined by its manufacturer and stated on each unit or in its user's manual. The recovery rate is the amount of water that the unit can heat to a temperature of 100°F in one hour. No matter how it is listed, you have to add the recovery rate plus 70 percent of the tank capacity to compute how much hot water the unit can deliver in one hour.

Hot water heaters are always full of water. The moment hot water leaves the tank an equal amount of cold water flows into it. But cold water lowers the temperature of all the water in the tank and if 70 percent of the tank water is cold, the temperature of hot water leaving the unit will be less than the prescribed 140° to 180°F. Thus, if your heater has a recovery rate of 60 gallons per hour, and the tank holds 50 gallons, the actual amount of hot water that you can draw in an hour is 60 gallons, plus 70 percent of 50 gallons (35 gallons), or 95 gallons.

HOT WATER HEATERS

Several types of hot water heaters are used in homes. They can be either oil- or gas-fired or electrically or solar heated, but all of them function by containing several gallons of water in a heated

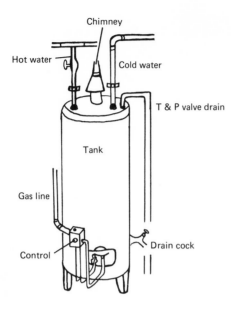

A gas-fired water heater.

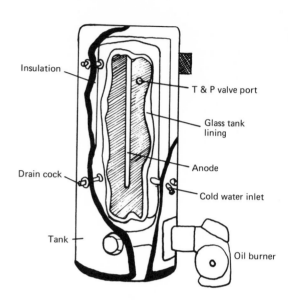

An oil-fired water heater.

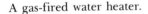

tank until that water attains a preselected temperature. The recommended temperature for hot water in a home is 140°F, although you may have to set the temperature slightly higher than that if it is servicing a dishwasher.

Every non-solar water heater consists of a metal tank, which in all recent models is glass-lined, and heavily insulated with fiberglass or rock wool; one or two heating elements; a pressure-relief valve; and a thermostat that controls the water temperature. Many heaters have a magnesium rod that hangs down in the tank. The rod, or anode, is used to attract any damaging chemical action away from the walls of the tank so that they will not corrode. The bottom of the tank normally has a drainage spigot. The tank should also be equipped with a temperature/pressure relief valve. The temperature and pressure valve is designed so that if the water temperature gets too high, it will automatically bleed off excess pressure. Or if the pressure gets too high, that will bleed off, too. Modern heaters also incorporate an energy cutoff device that closes down whatever fuel supply the unit uses any time the water temperature exceeds 200°F.

The heaters that use natural gas contain a flue area that runs through, or around, the appliance and is vented outdoors to siphon off fumes from the gas flame that burns under the storage tank. When the heater's thermostat calls for heat, a pilot flame ignites the burner, which burns until the water reaches the desired temperature; then the thermostat shuts off the gas supply. Modern gas burners are all designed so that should the pilot flame go out, the entire gas supply is automatically closed off to prevent any gas from escaping through the house.

Liquefied Petroleum (LP) gas heaters operate in the same way as their natural gas brethren. However, the fuel and heating orifices are smaller because the gas is more concentrated. The liquid gas is stored in a pressure tank attached to the heater. The gas vaporizes on leaving the storage tank, passes through a pressure reducer and goes to the heater, where it is combined with air and burned. LP-gas water heaters should have a 100 percent shutoff valve as a safety measure.

Oil heaters employ a pressure-fire burner similar to the kind used in oil furnaces, and have a combustion chamber under the water tank. Normally, the fuel that heats the water heater is piped directly from the furnace oil burner.

Electric water heaters may have either one or two heating elements. These are electrically insu-

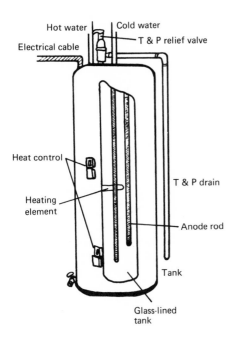

An electrically powered water heater.

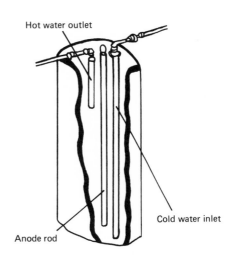

Plumbing connections to a water heater.

lated and immersed in the water. If there are two heating elements, each one is controlled by its own thermostat. Electric heaters usually operate on a 240v circuit and do not require a flue vented to the outside.

How Hot Water Heaters Work

Cold water is diverted directly from the cold water main to an inlet in the top of the heater tank and enters a dip pipe, which reaches almost to the bottom of the tank. This allows the new water to mix gradually with the warmed water in the tank so that it does not reduce the temperature in the tank too rapidly. As cold water flows into the tank, hot water simultaneously flows out the top of the tank through an outlet that actually is the beginning of the house hot water line that runs through the entire building. Any time hot water is drawn from the heater tank it is immediately replaced by cold water, which in turn is warmed to the proper temperature.

If the water in your locale is designated as hard, it will contain chemicals that will, in time, corrode the sides of the heater tank. Consequently,

modern units include a magnesium anode rod which extends down the center of the heater tank. The anode rod will sacrifice itself to corrosion so that is is eaten away instead of the tank. Still, whenever a water heater develops a leak it is usually because of corrosion.

Installing a Water Heater

Specific instructions for installing any modern water heater come with the given appliance, and they should be followed closely. No matter what model or type of heater you are installing, there are some general rules to remember:

Shut off the fuel supply at its source. That means shut down the furnace, turn off the main gas valve, remove the fuse in the electrical line, or turn off the circuit breaker.

Close the main water supply valve to the house. Open the nearest hot and cold water faucets to drain any water left in the pipes.

Be sure you have all the couplings and fittings you need to adapt the new heater to your plumbing system. Most of the fittings are supplied with the heater, but you should install a cutoff valve on the cold water supply if one is not already there.

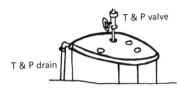

The temperature and pressure valve.

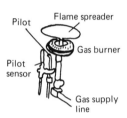

The burner in a gas-fired heater may look a little different from this, but it will include a pilot light, gas supply tube, and flame spreader.

You will also need a union for both the cold and hot water supply lines if not provided.

Temperature and Pressure Valve

Whether it came with the heater or not, install a temperature and pressure relief valve that is approved by your local plumbing code. It goes in a separate ¾-inch tapping in the tank. The purpose of a T and P valve is literally to keep the unit from exploding if all its power cutoffs fail at one time. All new heaters have a knockout in their tops so that you can easily install a T and P valve; the capacity of the T and P valve should be at least equal, or better still exceed, the heater's Btu-per-hour output. The valve is attached to a sensing probe that should extend at least 6 inches into the water at the top of the tank. The part of the valve that remains outside the tank is threaded to go into a ¾-inch tapping. Use Teflon tape on the threads.

When you install a T and P valve, it is inserted through the opening in the heater designated for it and then tightened with a wrench. A ¾-inch pipe is then threaded to the relief valve outlet and run to a suitable drainage within 6

inches of the floor. Do *not* put a cap on the pipe or a valve in it, and do not connect it to a drain. If you do, you will never know whether the valve is activated or not. You can put a pail under the open end of the hose, or just let the water drip on the floor. Chances are the valve will never have to be activated, but it is a piece of safety insurance you cannot risk being without.

Piping a Heater

When installing the piping for a hot water heater, be sure to install a shutoff valve in the incoming cold water line and make your connections with unions so that the tank can be removed and replaced without cutting out any pipes. Sometimes corrugated, flexible copper connectors are used on the hot and cold water lines to facilitate their hookup. If you want full hot water flow, all water heater plumbing should be done with ¾-inch pipe.

Connecting a gas- or oil-fired water heater to a CPVC or PB water supply system requires one special precaution to protect the plastic piping from conducted burner heat. Install 8-inch to 11-inch galvanized steel or brass nipples in the hot and cold tappings atop the tank, then start your plastic plumbing with a pair of ¾-inch transition unions, which are adapters between the threaded and plastic pipe. CPVC will solvent-weld directly to the transitions; PB requires a CPVC/PB adapter, which solvent-welds to the transition. Of course, a threaded metal gate valve may be installed to the cold water nipple before adapting to plastic; or a plastic valve may be installed after adapting. In any case, the metal nipples will protect the plastic piping from conducted burner heat; they are designed to stand up to the hot water coming from the heater.

Anode

Hard water exists in many areas of the country, so modern heaters often include an anode rod. The anode is a long magnesium rod threaded into the top of the heater so that it extends almost to the bottom of the tank. Because magnesium corrodes more easily than steel, any electrolytic action in the water attacks the rod before it does the sides of the tank. A residual of this chemical reaction is that it also plates the walls of the tank as if they were silver-plated dinnerware and protects the steel as well as makes the water in the tank less corrosive. Anodes have a life expectancy of between 3 and 10 years. You should, nevertheless, pull the rod out of the tank once a year and examine it. If it is almost eaten away, replace it.

Water Tanks

Most of the newest heater tanks are glass-lined. But, corrosion can still take place in the tank plus the fact that many hard water areas have a problem with dirty water. If puddles of water show up under your heater, you can assume the tank has corroded through at its weakest point. The easiest, and almost the least expensive, recourse is to install a new heater.

Burners

Gas-fired burners vary somewhat but all of them consist of a flame spreader positioned over a gas supply pipe at the bottom of the unit, which is controlled by a thermostat attached to the outside of the tank. The gas burner has an air control slot that you can regulate by turning a set screw. There is also a pilot light tube and pilot sensor. On very rare occasions you may have to shut off the gas line and clean the burner holes with a toothpick or needle.

Electrical Requirements

Electric power requirements for electric water heaters vary with each model. But as a rule, if they need 3,500 watts, they require a 12-gauge wire and 20-ampere fuse; 4,500 watts need a 10-gauge wire and 30-ampere fuse; 5,500 watts require an 8-gauge wire and 40-ampere fuse. All electric heaters should be on 240-volt AC lines except for those of 30 gallons or less, which can be operated on a 120-volt line.

If there are two heating elements, one is positioned behind an access panel at the bottom of the appliance and the other is near the top. They may be immersed in the water or wrapped around the tank. Both types are easily replaced, should they fail. The user's manual that accompanies the heater probably explains the necessary procedures for repairing the electrical elements in the heater.

Note: Never turn on the power to any hot water heater unless the tank is completely full of

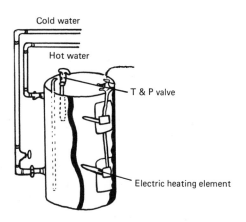

Cold water

Hot water

T & P valve

Electric heating element

Be sure there is water in the heater before turning on the power to an electric water heater.

water. The heating element(s) will burn out very quickly.

AVOIDING TROUBLE WITH WATER HEATERS

Water heaters are built to operate for years without developing any trouble. But like any appliance, they will remain trouble-free even longer if you give them a few minutes of attention from time to time:

1. Pull the anode out of the tank each year to see if it should be replaced.

2. If your area has very hard water, drain a pail or so of water out of the heater once a month. This will get rid of any sediment that may have settled on the bottom of the tank.

3. Lift the T and P valve test lever once a month. If it gives off water, the valve is working properly. If nothing happens, consider replacing the valve.

4. Every gas-fired or oil-fired heater must have a flue vented out of doors. Hold a lighted match in the draft diverter or draft regulator opening while the unit is firing. If the flame is drawn into the opening, the flue is in working order. If the flame is not drawn into the flue, clean the flue. Gas-fired or oil-fired heaters should also be cleaned and adjusted every two years.

THE TROUBLES WITH WATER HEATERS

Of the few times a water heater does develop problems, only a few instances can be solved by plumbing. Most of the times a heater breaks down, the problem is likely to center around the heating element or its thermostat. Because the electrically powered units have the most components they are also likely to have the most problems. It is beyond the scope of this book to detail the nonplumbing repairs that might be made on a water heater. However, some of the possible causes are listed in the checklists that follow so that you will have some idea of where the trouble may reside.

WATER HEATER REPAIR CHECKLIST

Problem: Water too hot or too cold

Possible Cause	Solution
Thermostat improperly set	Reset the thermostat to 140°–160°F ("warm" to "medium").
T and P valve defective	Lift the valve lever manually. If no water comes from the valve, replace it.

Problem: Faulty thermostat

This probably requires replacement of the thermostat.

Problem: Defective heating element

If the heater is oil- or gas-fired, try cleaning the heating element. If the heater is electric, the element will most likely have to be replaced.

Problem: High water heating costs

Possible Cause	Solution
Leaking faucets	Inspect the washers and packing in all faucets and replace any that are worn or chewed.
Thermostat set too high	Reset the thermostat as low as you can without too frequently running out of hot water.
Heater element seal leaks	Clean and check the seal(s) for damage and replace if they are faulty.
Thinly insulated water tank	Install an add-on insulation kit.
High piping heat losses	Insulate the house hot water pipes.
Heavy hot water loss	Install water-saving faucet and shower devices.

(Continued)

WATER HEATER REPAIR CHECKLIST (Continued)

Problem: Leaking water tank

Possible Cause	Solution
Leaking tank	Replace heater.
Leaks at plumbing connections	Reconnect plumbing with watertight joints.
Leakage around heating element (electric unit)	Replace the element gasket.
Leakage around thermostat	Replace thermostat gasket.

Problem: Little or no hot water

Possible Cause	Solution
Not enough power supply	Check the power supply to be sure the unit is getting enough gas, oil, or electricity.
Thermostat not adjusted	Adjust the thermostat to 140°–160°F.

Problem: Incorrect wiring (electric unit)

Have an electrician look at it.

Problem: Defective heating element

Install a new element. (Disconnect all the heater's power first.)

Problem: Heater too small for family needs

Replace the heater with a proper sized unit, or add a solar collector to its cold water supply.

Problem: Steam in water (Dangerous)

Possible Cause	Solution
Faulty thermostat or burner	Replace or have serviced. Check T and P valve. Replace if not functioning.
Grounded heating element (electric unit)	Replace element. (Disconnect heater's power supply first.)

Problem: Heater recovery rate slow

Possible Cause	Solution
Thermostat defective	Replace.
Top heating element burned out (electric unit only)	Replace. (See precaution above.)

Problem: Water dirty

Possible Cause	Solution
Mud or silt in tank	This can happen when a municipality turns on all the hydrants in a given area. Turn off the power supply and drain the tank, then refill it and turn on the power.

(Continued)

WATER HEATER REPAIR CHECKLIST (Continued)

Problem: Heater hisses when turned on

Possible Cause	*Solution*
Leaking tank	Replace the heater.
Leaking pipe or connection	Inspect the pipes leading into and out of the heater for leakage. Repair any leaks.

Problem: T and P valve pops continuously

Possible Cause	*Solution*
Temperature set too high	Reset the thermostat to 140°–160°F.
Faulty relief valve	Replace the T and P valve.
Excessive water pressure	Install a house pressure-reducing valve.

Glossary

Angle valve. Any globe valve having its inlet and outlet ports at right angles to each other.

Antisiphon trap. Any trap designed to prevent the siphoning of water through it.

Back siphoning. The flow of water into the water supply system from any source other than the source of water supply. Also known as back flow.

Back vent. The portion of the vent line that attaches directly to a trap under a fixture and extends to a waste pipe. Also known as an individual vent.

Backwater valve. A type of check valve used to prevent backflow (back siphoning) into a basement.

Bag trap. A trap shaped like a bag.

Ball cock. A faucet that is opened or closed by the rising or falling of a floating ball inside the faucet body.

Ball valve. A valve that controls the flow of water through it via a small ball that is rotated by the valve handle.

Basket strainer. A screen-type fitting placed in a drain hole in the bottom of a kitchen sink.

Bell (hub). The end of a pipe that is large enough to contain the end of a straight pipe of the same diameter.

Branch. Any part of a piping system other than the main line.

Branch vent. A vent pipe connecting the branch of a drainage system to the vent stack.

Building sanitary drain. Any building drain that conveys only sewage.

Building drain. The lowest drainage pipe in a house drain system. All other drains are connected to the building drain, which is connected to the building sewer outside the house.

Building drain branch. A waste pipe that extends horizontally from the building drain to service fixtures.

Building sewer. A drainpipe leading from a building to a public sewer. The building sewer exits entirely outside a building.

Building storm drain. A drain that carries only storm (rain) water.

Building supply line. The pipe entering the house from the street that carries potable water into the building.

Butterfly valve. A valve having a rotating disk that opens or closes the valve by rotating its handles 90°.

Bypass. Any method that allows water to pass around a valve or fixture connection.

Check valve. A valve that permits the flow of water in only one direction. Check valves close automatically to stop any water from returning through the pipe.

Cleanout. A threaded plate that plugs the ends of drainage fittings, which can be removed to allow cleaning of the drain line.

Collar. A sleeve attached to the back of a flange.

Common seal trap. A P-trap having a curve that is 2″ to 4″ deep.

Common vent. A vent connecting two fixture drains and serving both drains.

Compression faucet. A faucet that closes off the flow of water by means of a seat washer that is compressed over its seat.

Compression stop. A nonrated globe valve.

Continuous vent. Any vent that is a continuation of a drainpipe. The upper portion of a stack, for example, is a continuous vent from the highest fixture connected to the stack to where the pipe ends above the house roof.

Crown of a trap. The point in a trap where the direction of waste flow turns downward.

Curb cock. A valve placed in a house water service near the street.

Deep seat trap. A P-trap that is a 4″ in diameter having 1½″ inlet and outlet pipes.

Developed length. The length of a pipe including its fittings.

Dip of a trap. The point in a trap where the direction of waste flow turns upward.

Drain. A sewer or other pipe used for carrying sewage or waste water, including ground water, surface water, or storm water. **Drum trap.** A trap consisting of a cylinder that is installed vertically. The cylinder is normally 4″ in diameter having 1½″ inlet and outlet pipes.

Dry vent. A vent that does not carry water or water-borne wastes.

DWV. Abbreviation for Drain—waste—vent.

Female thread. A thread on the inside of a pipe or fitting.

Finishing work. Work done after the roughing in is complete. Normally, finishing work includes installation of the plumbing fixtures.

Fittings. Couplings, tees, elbows, or any of the connecting pieces used to hold lengths of pipe together.

Fixture branch. The supply pipe between a fixture and a water-distributing pipe.

Flush bushing. A bushing without a shoulder that fits flush into the fitting it connects to.

Fixture drain. The drain from any fixture trap, connecting to the drain pipe.

Flood level rim. The top edge of a fixture. When water reaches the rim it overflows.

Floor drain. A drainpipe that ends in a floor to allow water on the floor, to flow into the drainage system.

Flow pressure. The pressure of water in the supply lines.

Flow rate. The volume of water used by a plumbing fixture within a specific amount of time. Normally expressed in gallons per minute.

Flush. To wash out something with water.

Flush valve. A device placed at the bottom of tanks and toilets to permit them to be emptied.

Gate valve. A valve having a wedge-shaped disk that fits over seats within the body of the valve.

Globe valve. A valve that controls the flow of water by means of a disk that is fitted into a seat.

Grade (pitch). The slope of a pipe from its horizontal plane.

Group vent. A branch vent that services two or more traps.

Header. A pipe containing many outlets, to which other pipes are connected.

Horizontal branch drain. A drainpipe that extends horizontally from the stack to a fixture.

Horizontal pipe. Any pipe that lies at any angle less than 45° with the horizontal plane.

House drain. The lowest horizontal drainpipe in a building. It connects to all of the vertical drains and the outside sewer line.

Hub. The bell end of a cast-iron or plastic pipe (see Bell).

Increaser. A coupling that has one end larger in diameter than the other so that pipes of different diameters can be united.

Indirect waste pipe. Any waste pipe that does not connect directly with the drainage system.

Individual vent. A pipe used to vent a single fixture trap. It may connect to other vents, stacks, or go outside the house.

Invert. The lowest point on the interior of a horizontal pipe.

Lavatory. A fixture used for washing, found in bathrooms.

Main vent. The largest vent pipe in a system

(usually the top portion of the stack); used to vent the system.

Male thread. A thread on the outside of a fitting or pipe.

Offset. Any combination of elbows or bends that brings a pipe out of its normal line.

Pipe. A cylindrical conductor of water or waste fluids.

Plug. A pipe fitting used for closing and opening in other fittings.

Plumbing. The techniques of installing piping and fittings to bring water into, and carry water out of, a building.

Plumbing fixture. A receptacle for wastes and water that are released into the drainage system of the house.

Plumbing system. The plumbing system of a house includes all water supply pipes and all drainage pipes and their fittings, as well as all plumbing fixtures connected to the system.

Potable water. Drinkable water. To be potable, water must be freed of all impurities harmful to human beings.

Pressure-reducing valve. An automatic valve capable of reducing the amount of pressure of water entering the plumbing system.

Primary branch. The major branch of a building drain, which slopes from the base of the stack to the building drain.

P-trap. A P-shaped pipe used to contain water in the bottom of its curve, to prevent gases and vermin from entering the fixture it is attached to, as well as to catch debris that is large enough to clog other parts of the drain line.

Relief valve. A device that automatically protects against excessive temperatures or pressures.

Relief vent. A vent used to provide additional air circulation between the drain and vent systems.

Riser. Any pipe that extends vertically through a house. Principally refers to water supply pipes.

Roughing-in. The installation of all parts of the plumbing system up to the point of installing the fixtures.

Run. The continuing straight line of a pipe and its fittings.

Sanitary drainage pipe. Pipes used to remove waste from fixtures.

Sanitary drainage and vent piping system. Pipes installed to remove waste from a house and provide air circulation in the drainage pipes.

Sewage. Any liquid waste, often containing animal or vegetable matter.

Sewer. A large pipe that carries waste fluids only.

Sewer gas. A mixture of vapors, odors, and gases found in sewage.

Sill cock. A lawn faucet installed on the exterior of a building.

Soil pipe. A pipe used to convey sewage from water closets.

Soil stack. A vertical pipe extending from the house drain to the room of a house that connects to various fixture drains.

Stack. A general term referring to any vertical soil pipe.

Stack cleanout. A plugged fitting found at the base of all stacks.

Stack group. The fixtures located near, and connected to, a single stack.

Stack vent. The extension of the soil stack from the topmost drain fixture connection to where the pipe ends above the roof.

Storm sewer. A sewer used to carry sewage as well as run-off water from the ground or surface.

S-trap. An S-shaped pipe installed under a fixture in the drain line to protect the fixture from gases and vermin.

Swing joint. A joint in a threaded pipe permitting the pipes to swing out of their normal line.

Temperature and relief valve. A safety valve designed to release pressure in a pipe system or heating device should the temperature of the water become too hot.

Trap. A fitting that provides a liquid seal to prevent gases and vermin from entering a drainage system.

Tube. A thin-walled conduit used for carrying potable water in a water supply system instead of pipes. Tubing can be bent, while pipe cannot.

Vacuum. Any air pressure that is less than the pressure exerted by the atmosphere.

Valve. A device used to control the flow of water.

Vent system. A system of pipes used to provide an air flow through a drainage system to protect trap seals from back pressure.

Waste/overflow fitting. A bathtub drain fitting that controls both the tub drain and any overflow from the tub.

Waste pipe. A pipe that carries only liquid waste.

Waste stack. A vertical pipe that carries waste from one or more fixtures to the house drain.

Water closet. Technical name for a toilet.

Water hammer. The rattling of pipes every time a faucet is closed.

Water heater. A device used to heat water to a temperature of 120°F to 160°F.

Water meter. A device used to measure the amount of water consumed by a household.

Water service. The pipe from a water main to the water supply system in a house.

Water softener. An appliance that chemically removes impurities from potable water.

Water supply system. The water service pipes throughout a house running from the service entrance to each fixture.

Wet vent. Any vent pipe that also serves as a drain.

Index